4220903123427O

W0115690

STUDENT WORKBOOK

to accompany

PROGRAMMING OF
CNC MACHINES

Fourth Edition

Ken Evans

Industrial Press, Inc.

Industrial Press, Inc.
32 Haviland Street, Suite 3
South Norwalk, Connecticut 06854
Tel: 203-956-5593, Toll-Free: 888-528-7852
E-mail: info@industrialpress.com

Programming of CNC Machines, Student Workbook, 4th Edition, by Ken Evans.

ISBN: 978-0-8311-3600-0
ISBN ePDF: 978-0-8311-9380-5
ISBN ePUB: 978-0-8311-9381-2
ISBN eMOBI: 978-0-8311-9382-9

Copyright © 2016 by Industrial Press, Inc.
All rights reserved.

This book, or any parts thereof, with the exception of those figures in the public domain,
may not be reproduced, stored in a retrieval system, or transmitted in any form
without the permission of the copyright holders.

Sponsoring Editor: Jim Dodd
Developmental Editor: Robert Weinstein
Interior Text and Cover Designer: Janet Romano-Murray

Notice to the reader: While every possible effort has been made to insure the accuracy of the information presented herein, the authors and publisher express no guarantee of the same. The authors and publisher do not offer any warrant or guarantee that omissions or errors have not occurred and may not be held liable for any damages resulting from the use of this text by the readers. The readers accept the full responsibility for their own safety and that of the equipment used in connection with the instructions in this text. There has been no attempt to cover all controllers or machine types used in the industry and the reader should consult the operation and programming manuals of the machines they are using before any operation or programming is attempted.

industrialpress.com
ebooks.industrialpress.com

10 9 8 7 6 5 4 3 2 1

Table of Contents

Acknowledgments

I give thanks first to the Lord, Our God, for blessing me with the opportunity, knowledge, and ability to share in this work. Many thanks are due to all of the parties listed below, who helped on this project. Special thanks are due to the publisher, Industrial Press, including President Alex Luchars; Managing Editor, Laura Brengelman; former Editorial Director, John Carleo; and Production Manager, Janet Romano.

Thanks to Robert Weinstein, Editor, of Gerson Publishing Company, for his efforts in editing this text.

Thanks as well to:

- T.J. Long; Katie Richardson, and Larry Meenan of Kennametal for their assistance and contribution of tooling graphics and other technical data from the NOVO online application.
- Greg Mercurio of Shop Floor Automations for allowing the use of Predator Virtual CNC software for program verification in the previous edition.
- Carr Lane Manufacturing for technical data charts.

About the Author

Ken Evans has held diverse machining and related jobs throughout his career and is currently a CNC Programmer at a prominent aerospace company. He learned the machinist trade in 1976 at Cessna Aircraft in Wichita, Kansas. Evans began his formal teaching career in 1984 at the T.H. Pickens Technical Center in Colorado, while working full-time as a CNC machinist and quality control inspector.

From 1991 to 2010, he served as a Machine Tool Technology instructor at Davis Applied Technology College in Kaysville, Utah, teaching foundational through advanced-level courses in the machining curriculum, including Mastercam CAD/CAM classes for students, educators, and private industry. He also was designated a certified Project Lead the Way, Computer Integrated Manufacturing (CIM) instructor. From 1997 to 2003, Ken worked for a local machine tool distributor in Salt Lake City as a MAZAK certified Training and Applications Specialist and one of the nation's first Mazatrol Conversational Programming instructors.

Ken loves the outdoors; he enjoys gardening, mountain biking, and golf.

Preface

Many textbooks written on the subject of CNC programming include a multitude of practical examples; however, very few include enough practice exercises for readers to verify their understanding. This edition of *Student Workbook to Accompany Programming of CNC Machines* provides many practical exercises designed to verify comprehension of CNC programming. Although this workbook is written as a companion to *Programming of CNC Machines, Fourth Edition*, it may be used to confirm anyone's knowledge of CNC programming. There are many ways to program a part in order to get the accurate results. The proven method offered in the answer key is consistent with the companion text. Most important, the part is produced to specifications, both safely and efficiently.

It is possible for someone to program a CNC machine without actual machining experience. Still, the result will be better if a practical background exists. Feeds and speeds, tool selection, and work holding methods are very hard to learn from a book. Practical experience has been proven to be the best teacher. It is also a plus to have a thorough understanding of shop mathematics. A student who has been taught the basic fundamentals in these matters, beforehand, will have the most success writing CNC programs in the long term.

Units 1 through 4 of this workbook should be used to verify that you have learned the basic skills necessary to write CNC programs, line-by-line, to make a variety of workpieces. The answer key for each unit provides answers/solutions to verify correct programming. Try to complete as many questions and examples as possible without using the answer key. It would be better to use the text (*Programming of CNC Machines, Fourth Edition*) to look up the required information, rather than go directly to the answers. The text is a reference *tool*; just like your machinist tools, it was designed to help you get the job done. Whenever it is possible, consult with your trainer for additional ideas or methods to evaluate your work.

STUDENT WORKBOOK

to accompany

PROGRAMMING OF
CNC MACHINES

Fourth Edition

Unit 1: CNC Basics

Process Planning

Any time a new part is considered for manufacture, it is necessary to have a logical plan in order to machine it efficiently. The following is an explanation of the exercise requirement: Three Process Planning Sheets are provided on the following pages for operations (Chart 1-1), setup (Chart 1-2), and quality control (Chart 1-3). Copy as many as you need. Later in the workbook, CNC Programming Sheets are included for the units covering Turning Center Programming and Machining Center Programming. For a detailed description of the use of these documents, please refer to *Programming of CNC Machines*, *Fourth Edition*, Part 1, CNC Basics. You will also find in the units covering Turning Centers and Machining Centers a list of cutting tools that can be used for preparing the Process Planning Sheets.

Use the Operation Sheet (Chart 1-1) to identify each individual operation and the machines necessary to complete the part in the blueprints that follow.

Use the CNC Setup Sheet (Chart 1-2) to identify work holding, cutting tools, work piece coordinate zero locations, and any other pertinent information needed to complete the part setup for these blueprints. Refer to the tool lists provided in Unit 3 (CNC Turning Center Programming) and Unit 4 (CNC Machining Center Programming) to choose the appropriate tools.

Use the Quality Control Check Sheet (Chart 1-3) to list 100% of the dimensional data needed to verify that the parts are made to specification, starting with the blueprints in Figures 1-1 and 1-2.

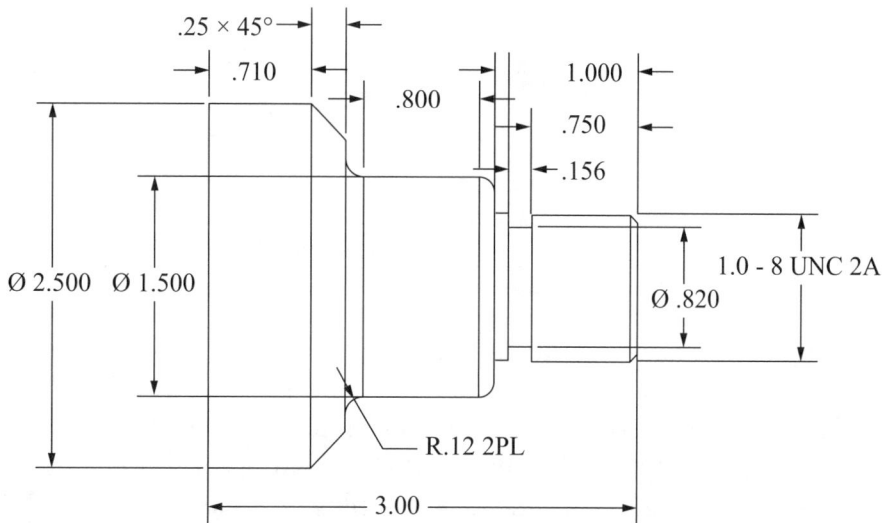

Figure 1-1   Turning Center Process Planning

1

Chart 1-1
Process Planning Operation Sheet

Date			Prepared By	
Part Name			Part Number	
Quantity			Sheet ___ of ___	
Material				
Raw Stock Size				
Operation Number	Machine Used	Description of Operation		Time

In the example in Figure 1-1, the parts are provided as 3.0625 long slugs with one end having already been faced. It will be necessary to clamp the 2.50 diameter in pre-machined soft jaws with enough material extended to allow machining of the part, including removal of 1/16 inch from the face of the part. The material is 4340 alloy steel. Dimensional tolerances are as follows: .X = plus or minus .015 inch, .XX = plus or minus .010 inch, .XXX = plus or minus .005 inch, and angular tolerance is plus or minus .5 degree. Please use copies of the Process Planning sheets and develop a plan to machine the part to dimensional requirements.

Chart 1-2
Process Planning CNC Setup Sheet

Date	Prepared By
Part Name	Part Number
Machine	Program Number

Workpiece Zero: X _____ Y _____ Z _____

Setup Description:

Tool Number	Offset Number	Tool Description	Comments

Chart 1-3
Process Planning Quality Control Check Sheet

Date			Checked By	
Part Name			**Part Number**	
			Sheet ___ of ___	
Blueprint Dimentsion	**Tolerance**	**Actual Dimension**	**Comments**	

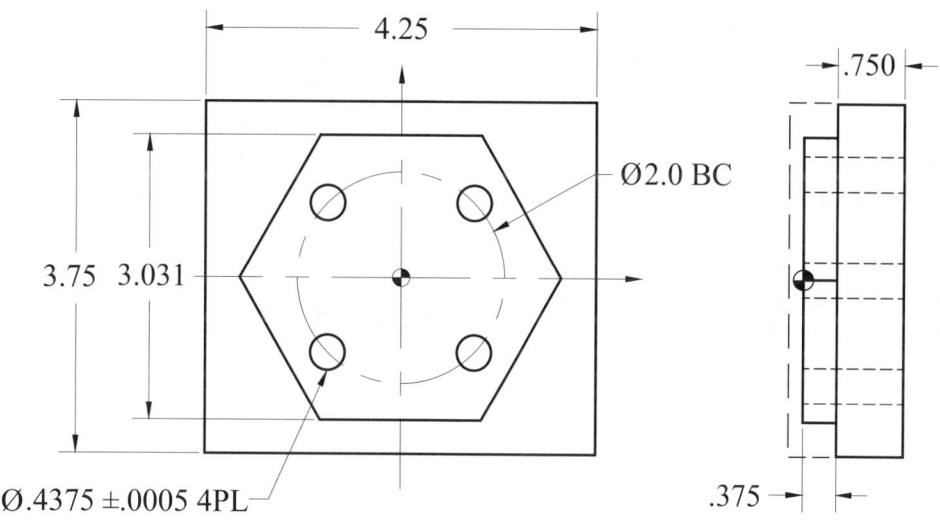

Figure 1-2   Machining Center Process Planning

In the example in Figure 1-2, it will be necessary to machine the part from a solid blank of Aluminum that is pre-machined to 4.25 square and is 1.25 thick. The top surface must have .125 inch of material removed as well. Dimensional tolerances are as follows: .X = plus or minus .015 inch, .XX = plus or minus .010 inch, .XXX = plus or minus .005 inch, and angular tolerance is plus or minus .5 degree. Please use the copies of the Process Planning sheets and develop a plan to machine the part to dimensional requirements. Make your tool selections from the CNC Machining Center Tool List found in Unit 4, CNC Machining Center Programming.

Chart 1-1
Process Planning Operation Sheet

Date			Prepared By	
Part Name			Part Number	
Quantity			Sheet ___ of ___	
Material				
Raw Stock Size				
Operation Number	Machine Used	Description of Operation		Time

Chart 1-2
Process Planning CNC Setup Sheet

Date	Prepared By
Part Name	Part Number
Machine	Program Number

Workpiece Zero: X _____   Y _____   Z _____

Setup Description:

Tool Number	Offset Number	Tool Description	Comments

Unit 1: CNC Student Workbook

Chart 1-3
Process Planning Quality Control Check Sheet

Date			Checked By	
Part Name			Part Number	
			Sheet ___ of ___	
Blueprint Dimentsion	**Tolerance**	**Actual Dimension**	**Comments**	

Feeds and Speeds

Charts 1-4 and 1-5 are supplied for use in answering the exercise problems presented here. The Surface Feet per Minute (SFPM) and Feed in inches per revolution (in/rev) values represented here are given in ranges. Note that the values starting at the low end of the range are intended as a minimum starting point for calculations; those at the high end are a maximum recommended SFPM and Feed. Final values used for machining may differ based on many factors. As you answer the problems, use values that are within the ranges given. Refer to *Programming of CNC Machines* (Part 1, CNC Basics, Metal Cutting Factors) for detailed information regarding Feed and Speed calculations. A more comprehensive source for machining data is the *Machinery's Handbook*; other valuable sources for machining data are in the tool and insert catalogs and online applications supplied by cutting tool manufacturers.

Chart 1-4 Feeds and Speeds for Turning

	Tool Material	
Material	**High Speed Steel**	**Carbide**
Carbon Steel **Feed in/rev**	30–60 .006–.012	200–1300 .008–.036
Alloy Steel **Feed in/rev**	30–120 .006–.012	125–1000 .008–.036
Stainless Steel **Feed in/rev**	25–110 .006–.012	80–945 .007–.036
Aluminum **Feed in/rev**	500–800 .006–.012	2800–4500 .017–.036

Note: As a general rule, the minimum depth of cut should be 1.5 to 2 times the tool nose radius. The maximum feed rate should be approximately one half the tool nose radius for rough turning using carbide inserts.

For milling, the maximum depth of cut is equal to the flute length or the insert height and the maximum width of cut is the cutter diameter. However, these measures are not practical in most cases. A more widely used practice is to set the maximum depth of cut to 2/3 of the flute length and the maximum width of cut to 2/3 of the diameter, as well. These basic conditions should be followed for the remainder of this workbook. Drilling calculations should be based on High Speed Steel (HSS) values for Turning and HSS End Mill values for Milling.

Chart 1-5 Feeds and Speeds for Milling

	HSS End Mill	Carbide End Mill	Carbide Inserted Face Mill
Carbon Steel **Feed in/tooth**	25–140 .001–.004	210–1000 .006–.012	90–685 .020–.039
Alloy Steel **Feed in/tooth**	5–85 .001–.004	40–450 .006–.012	39–475 .020–.039
Stainless Steel **Feed in/tooth**	20–80 .001–.003	200–700 .006–.012	210–385 .020–.039
Aluminum **Feed in/tooth**	165–850 .002–.006	600–2000 .008–.015	755–1720 .020–.039

Refer to the following formula to calculate revolutions per minute (r/min).

$$r\,/\min = \frac{12 \times CS}{\pi \times D}$$

where

CS = Cutting Speed from the charts above or the *Machinery's Handbook*

π = 3.1417

D = Diameter of the workpiece or the cutter

Refer to the charts above or the *Machinery's Handbook* for the feed in inches per tooth (in/tooth) for chip load recommendations. Also review the formula below to calculate the feed aspect of the metal-cutting operation.

$$F = R \times N \times f$$

where

F = Feed rate in inches per minute (in/min)

R = r/min calculated from the preceding formula

N = the number of cutting edges

f = the chip load, per tooth, recommended from the charts above or the *Machinery's Handbook*

1. On a CNC lathe, a facing cut is needed to establish the part-zero surface. The alloy steel bar stock is 2.5 inches in diameter and has 1/32 inch of excess material to be removed from each side. A carbide-inserted tool with a 1/32 inch nose radius will be used for this operation. Because the diameter changes as the tool travels toward the centerline, what should the r/min be? What should the SFPM be? What should the depth of cut be?

2. When finish turning an aluminum bar that is 2.3125 inches in diameter with a carbide-inserted turning tool that has a 1/64 inch tool nose radius, what is the r/min and feed rate required if the depth of cut is 1/64 inch per side?

3. An internal threading operation is required on a CNC lathe to make a 1-8 UNC thread in a carbon steel part. The cutting tool material is High Speed Steel. What should the r/min be for this operation?

4. Calculate the appropriate speeds and feeds for each of the required tools in the lathe process planning project above and enter your answers on your CNC Setup sheet in the comments section.

5. Calculate the appropriate speeds and feeds for each of the required tools in the mill process planning project above and enter your answers on your CNC Setup sheet in the comments section.

6. In this example, the material is stainless steel. A .5625 inch diameter hole is to be drilled through a plate that is 1.25 inch thick. Calculate the r/min and feed rate best suited for this operation. Use the HSS end mill values from the chart.

7. A carbon steel plate 4.0 inches square requires a 2.0 inch diameter hole to be machined through the center. A pre-drilling operation uses a 1.25 inch diameter HSS drill and a finishing operation uses a .875 diameter 4-fluted HSS end mill to circle mill out the remainder of material. What is the r/min and feed rate for the drill? What is the r/min and feed rate for the end mill?

8. A 5-tooth 3.0 inch diameter carbide face mill is used to machine an alloy steel bar that is 2.0 inches wide and 6.0 inches long. There are two depth passes of .080 inch each required to bring the part to size. What is the r/min and feed rate for this cut?

9. A 4.0 inch flat aluminum bar requires a profile to be cut on both ends. A 2-fluted HSS end mill 7/16 inch in diameter has been selected for the job. The part thickness is 1/2 inch and the amount of axial metal removal is 1/2 inch. What are the appropriate r/min and feed rate?

10. Use the formula and data given above to calculate the feed and speed required for each tool in the programming exercises that follow. List your results in the comments section of the CNC Setup Sheet.

Coordinate Systems

1. Use Figure 1-3 to identify the absolute coordinates for each axis and for each point of the profile of the turned part, based on diametrical considerations.

2. Use Figure 1-3 to identify the incremental coordinates for each axis and for each point of the profile of the turned part, based on radial considerations.

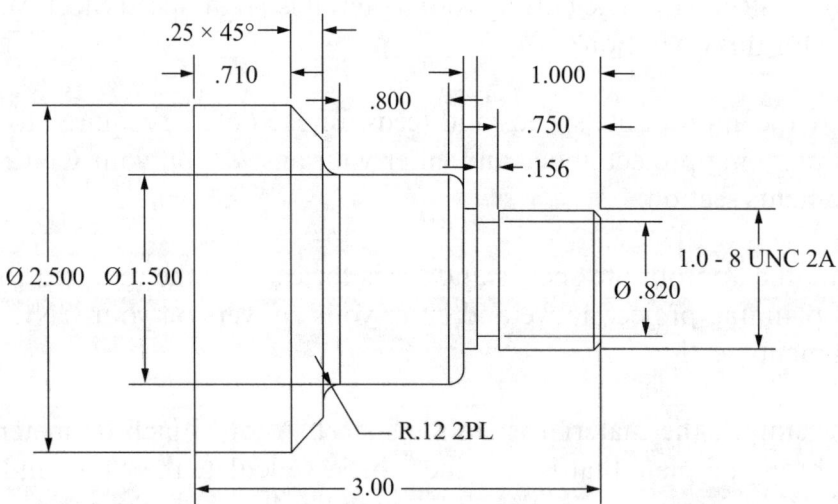

Figure 1-3   Identify Absolute and Incremental Coordinates

3. Use Figure 1-4 to identify the absolute coordinates for each axis and for each point of the profile of the milled part. Start at part zero and proceed clockwise.

4. Use Figure 1-4 to identify the incremental coordinates for each axis and for each point of the profile of the milled part.

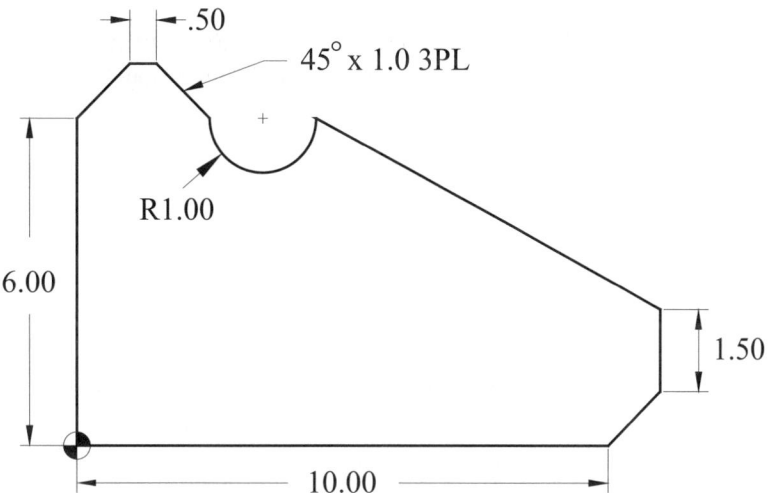

Figure 1-4   Identify Absolute and Incremental Coordinates

5. List the absolute coordinate values for X, Y, and Z for each of the 15 points as indicated in Figure 1-5). The part is 3.0 inches long, 2.0 inches wide and has a height of 2.25 inches. The slot is cut through the centerline of the width and is .50 wide and .375 deep.

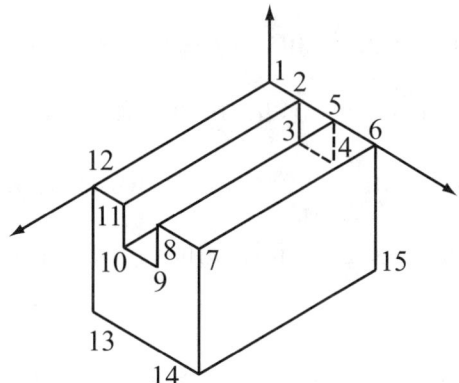

Figure 1-5   Identify Absolute Coordinates

6. Identify each axis (vertical milling representation) and its positive or negative value on Figure 1-6.

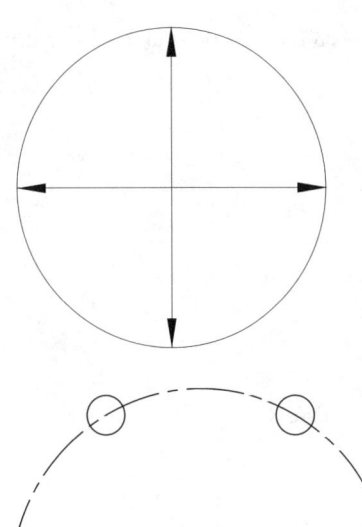

Figure 1-6   Identify Vertical Milling Axes, Polar Rotation, Quadrants, and Angular Values

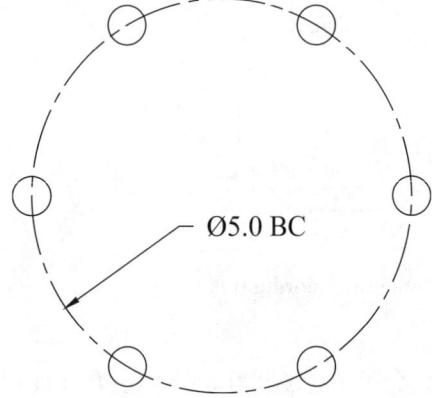

Ø5.0 BC

Figure 1-7   Identify Polar Coordinate Values

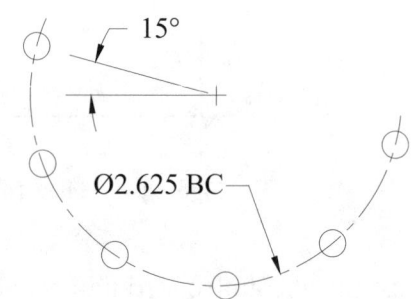

15°

Ø2.625 BC

Figure 1-8   Identify Polar Coordinate Values

7. Indicate the negative rotation direction for the polar coordinate system in Figure 1-6.

8. Indicate each of the polar quadrants in Figure 1-6.

9. Identify the angular value locations for 0, 90, 180, and 270 degrees in Figure 1-6.

10. Identify the polar (angular and radial) values for each of the holes in Figure 1-7.

11. Identify the polar (angular and radial) values for each of the holes in Figure 1-8.

Trigonometric Calculations

In many cases, it will be necessary to calculate the coordinate values of points for input into your CNC programs. The exercises that follow are a small sampling of the types of problems you are likely to encounter. Use your calculation skills to answer all of the problems and to prepare yourself for others when you complete the actual programming exercises in Units 3 and 4 of this workbook, the CNC Turning and Machining Center sections.

Chart 1-6 (Right Triangles) and Chart 1-7 (Oblique Triangles) are provided for your benefit. Many of the formulas shown are needed to complete the problems.

Chart 1-6 Right Triangles

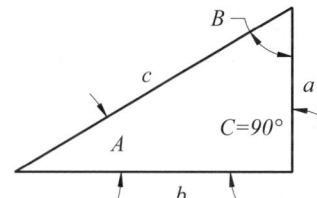

Known Sides and Angles	Unknown Sides and Angles			Area
a and b	$c = \sqrt{a^2 + b^2}$	$A = \arctan\dfrac{a}{b}$	$B = \arctan\dfrac{b}{a}$	$\dfrac{a \times b}{2}$
a and c	$b = \sqrt{c^2 - a^2}$	$A = \arcsin\dfrac{a}{c}$	$A = \arccos\dfrac{a}{c}$	$\dfrac{a \times \sqrt{c^2 - a^2}}{2}$
b and c	$a = \sqrt{c^2 - b^2}$	$A = \arccos\dfrac{b}{c}$	$B = \arcsin\dfrac{b}{c}$	$\dfrac{b \times \sqrt{c^2 - b^2}}{2}$
a and $\angle A$	$b = \dfrac{a}{\tan A}$	$c = \dfrac{a}{\sin A}$	$B = 90° - A$	$\dfrac{a^2}{2 \times \tan A}$
a and $\angle B$	$b = a \times \tan B$	$c = \dfrac{a}{\cos B}$	$A = 90° - B$	$\dfrac{a^2 \times \tan B}{2}$
b and $\angle A$	$a = b \times \tan A$	$c = \dfrac{b}{\cos A}$	$B = 90° - A$	$\dfrac{b^2 \times \tan A}{2}$
b and $\angle B$	$a = \dfrac{b}{\tan B}$	$c = \dfrac{b}{\sin B}$	$A = 90° - B$	$\dfrac{b^2}{2 \times \tan B}$
c and $\angle A$	$a = c \times \sin A$	$b = c \times \cos A$	$B = 90° - A$	$c^2 \times \sin A \times \cos$
c and $\angle B$	$a = c \times \cos B$	$b = c \times \sin B$	$A = 90° - B$	$c^2 \times \sin B \times \cos$

Chart 1-7 Oblique Triangles

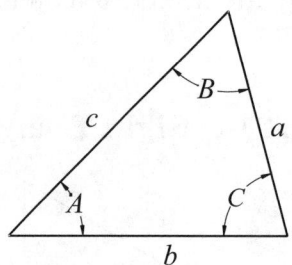

Known Sides and Angles	Unknown Sides and Angles			Area
All three sides a, b, c	$A =$ $\arccos\dfrac{b^2 + c^2 - a^2}{2bc}$	$B =$ $\arcsin\dfrac{b \times \sin A}{a}$	$C =$ $180° - A - B$	$\dfrac{a \times b \times \sin C}{2}$
Two sides and the angle between them a, b, $\angle C$	$c =$ $\sqrt{a^2 + b^2 - (2ab \times \cos C)}$	$A =$ $\arctan\dfrac{a \times \sin C}{b - (a \times \cos C)}$	$B =$ $180° - A - C$	$\dfrac{a \times b \times \sin C}{2}$
Two sides and the angle opposite one of the sides a, b, $\angle A$ ($\angle B$ less than 90°)	$B = \arcsin\dfrac{b \times \sin A}{a}$	$C = 180° - A - B$	$c = \dfrac{a \times \sin C}{\sin A}$	$\dfrac{a \times b \times \sin C}{2}$
Two sides and the angle opposite one of the sides a, b, $\angle A$ ($\angle B$ greater than 90°)	$B =$ $180° -$ $\arcsin\dfrac{b \times \sin A}{a}$	$C = 180° - A - B$	$c = \dfrac{a \times \sin C}{\sin A}$	$\dfrac{a \times b \times \sin C}{2}$
One side and two angles a, $\angle A$, $\angle B$	$b = \dfrac{a \times \sin B}{\sin A}$	$C = 180° - A - B$	$c = \dfrac{a \times \sin C}{\sin A}$	$\dfrac{a \times b \times \sin C}{2}$

1. In order to program the part shown in Figure 1-9, it will be necessary to identify the absolute rectangular coordinate location of the center point for each hole. List the coordinate values for each, starting with the hole at the one o'clock position and proceeding clockwise. The angular value for this hole is 70 degrees.

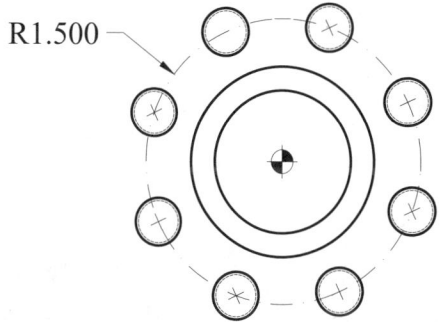

Figure 1-9   Calculate Absolute Coordinates

2. In order to inspect the part in Figure 1-10 to specification, a center-to-center dimension is required. Use the data given to calculate this dimension.

3. Use Figure 1-11 to calculate the values for the unknown distance.

4. Use Figure 1-12 to calculate the values for each chord distance.

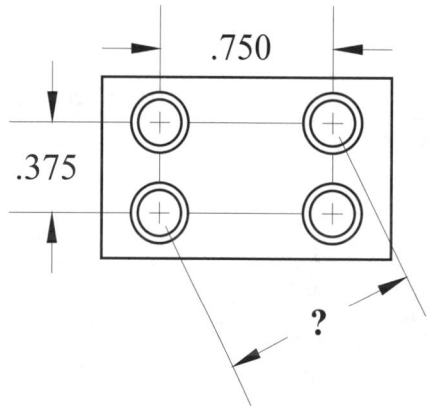

5. Calculate the amount of tool travel necessary, allowing for the drill point, to drill through a .562 inch thick plate using the drill diameter shown in Figure 1-13, with a standard drill point angle of 118°.

Figure 1-10   Calculate the Center-to-Center Distance

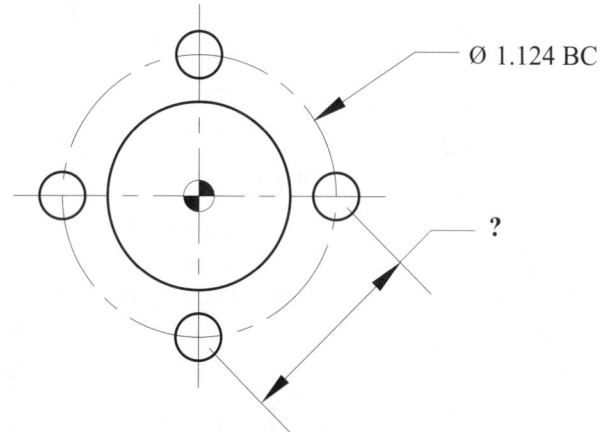

Figure 1-11   Calculate the Center-to-Center Distance

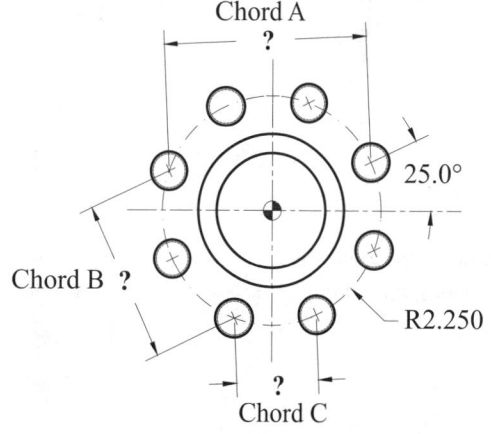

Figure 1-12   Calculate Chord Distances

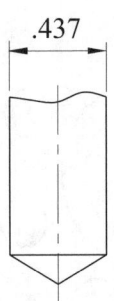

.437

Figure 1-13
Calculate for Drill
Point Compensation

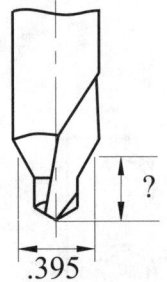

.395

Figure 1-14
Calculate for Drill Point
Compensation

6. Calculate the depth of cut required using a #5 (Plain Type) center drill to countersink to a diameter of .395 inch (Figure 1-14). Plain type center drills have an angle of 60° with a point angle of 120°. The length from the end of the point angle to the beginning of the 60° angle is 3/16 inch (see *Machinery's Handbook*).

7. Calculate the amount of tool travel necessary, allowing for the drill point plus .090, to drill through a .875 thick plate using the drill diameter shown in Figure 1-15 and with a drill point angle of 135°.

8. The profile of a part is to be machined using a .500 inch diameter end mill, as shown in Figure 1-16. Calculate the necessary offset amount for each axis and the coordinate values that will be required for the CNC program.

9. In Figure 1-17, a calculation is necessary to offset for the tool nose radius when turning a 30° tapered surface. The face and centerline of the turned part are zero. List the coordinates needed in the CNC program to allow for this offset.

.453

Figure 1-15
Calculate for Drill Point
Compensation

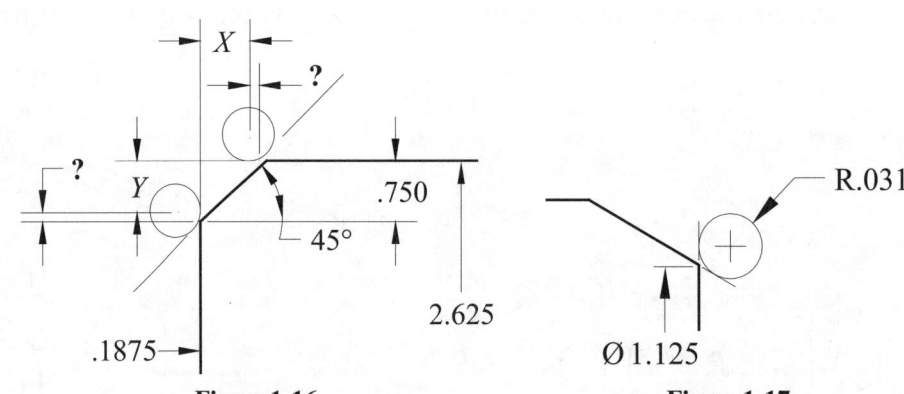

Figure 1-16
Calculate for Cutter
Offset Coordinates

Figure 1-17
Calculate for Tool Nose
Radius Offset Coordinates

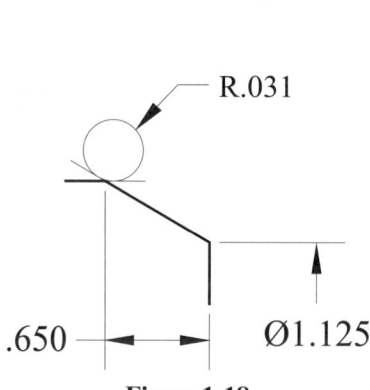

Figure 1-18
Calculate for Tool Nose
Radius Offset Coordinates

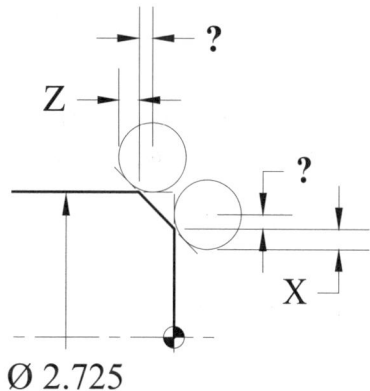

Figure 1-19
Calculate for Tool Nose
Radius Offset Coordinates

10. In Figure 1-18, a calculation is necessary to offset for the tool nose radius when turning a 30° tapered surface. The face and centerline of the turned part are zero. List the coordinates needed in the CNC program to allow for this offset.

11. In Figure 1-19, a calculation is necessary to offset the amount for the tool nose radius when turning a .062 inch 45° chamfer. In this case, the tool nose radius is .015 inch. List the coordinates needed in the CNC program to allow for this offset.

Study Questions: CNC Basics

1. Programming is a method of defining tool movements through the application of numbers and corresponding coded letter symbols.
   T or F

2. A lathe has the following axes:
   a. X, Y, and Z
   b. X and Y only
   c. X and Z only
   d. Y and Z only

3. Program coordinates that are based on a fixed origin are called:
   a. Incremental
   b. Absolute
   c. Relative
   d. Polar

4. On a two axis turning center, the diameter controlling axis is:
   a. B
   b. A
   c. X
   d. Z

5. The letter addresses used to identify axes of rotation are:
   a. U, V, and W
   b. X, Y, and Z
   c. A, Z, and X
   d. A, B, and C

6. The acronym TLO stands for:
   a. Tool Length Offset
   b. Total Length Offset
   c. Taper Length Offset
   d. Time Length Offset

7. When referring to the polar coordinate system, the clockwise rotation direction has a positive value.
   T or F

8. In the text *Programming of CNC Machines*, Figure 1-17 of Part 1, in which quadrant is the part placed?

9. A program block is a single line of code followed by an end-of-block character.
   T or F

10. Each block contains one or more program words.
    T or F

11. Using Figure 1-15, Part 1, *Programming of CNC Machines*, list the X and Y absolute coordinates for the part profile where workpiece zero is at the lower left corner. (The corner cutoff is at a 45° angle.) Use a clockwise rotation direction.

12. Using Figure 1-15, in Part 1, *Programming of CNC Machines*, list the X and Y incremental coordinates for the part profile where workpiece zero is at the lower left corner. Use a clockwise rotation direction.

13. How often should machine lubrication levels be checked?

Unit 2: Setup and Operation

The questions in this unit will require essay answers. List your answers in the space provided, or attach a separate sheet if more space is needed. Please consult with your instructor or the answer key at the end of this workbook to review and compare your solutions with the suggested answers.

General Steps

1. What are the general steps required in order to prepare a CNC machine for production of a programmed part that has been effectively run before?

Operation Scenarios

2. When using the Pulse Generator (Handle) to move a selected axis for setup purposes, there are several increments available. List the incremental step magnitude for each:

3. What will occur when the reset button is pressed during automatic operation? What steps should be followed to recover?

4. What differences would exist if the Emergency Stop button is pressed during automatic operation rather than the reset button? What will occur? What steps should be followed to recover?

5. When the Feed Hold button is pressed during automatic operation, what will occur? What steps should be followed to recover?

6. What mode of operation is required to install a tool into the spindle by the Automatic Tool Changer (ATC)?
    a. Automatic
    b. Jog
    c. Edit
    d. Manual Data Input (MDI)

7. It is highly advisable to conduct an in-process inspection of parts as they are machined for the first time. What is the appropriate and safe method for performing these measurements?

8. If a part is machined on the lathe by tool #4 and it is found that the outside diameter measurement is oversize by .003, where, and how is the change made to compensate for this variation?

9. If an existing wear offset is evident for tool #4 on the offset page of a lathe, what is the method for clearing this value so that no offset amount will remain?

10. In this case, the program execution has been interrupted. The tool that was being used is tool #6. How can the program be restarted from the beginning of that tool's use, and completed from that point on, rather that starting over at tool #1?

MDI Usage

11. Many times during the setup process, it is helpful to use Manual Data Input (MDI) to accomplish specific tasks. Name at least two of these tasks and explain how they would be useful. If possible, list the codes necessary to execute them.

Programming Editing

It is inevitable that program errors will happen and that program editing adjustments will be necessary. Answer the following questions with the procedures necessary to correct for this reality:

12. The program that is required calls for the G55 offset to be used on the CNC setup sheet. You have measured the offset data and entered it at that location on the offset page. When the program is loaded into active memory, it is noted on the program display that offset G54 is called in the program. Describe the course of action necessary to correct for this error.

13. The program you are using for the turning center does not include the tool or wear offset number for tool number 1. Describe the course of action necessary to insert the needed information into the program.

14. In the program, a comment (Date 03/30/15) is included. It is decided that it is not necessary to include this comment. Describe the course of action necessary to remove this portion of the program.

Study Questions: Setup and Operation

1. The counterclockwise direction of rotation is always a negative axis movement when referring to the handle (pulse generator).
   T or F

2. Which display includes the programmed Distance-to-Go readouts?

3. When the machine is ON and the program check screen is displayed, there is a list group of G-Codes displayed. What does this indicate?

4. Describe the difference between the Input and the +Input soft keys in the function.

5. Which button is used to activate automatic operation of a CNC program?
   a. Emergency Stop
   b. Cycle Stop
   c. Cycle Start
   d. Auto

6. Which display lists the CNC program?
   a. Position page
   b. Offset page
   c. Program check
   d. Program page

7. When the machine is turned on for the first time, it must be sent to its home position.
    T or F

8. Which operation selection button allows for the execution of a single CNC command?
    a. Dry run
    b. Single block
    c. Block delete
    d. Optional stop

9. Which mode switch/button enables the operator to make changes to the program?
    a. Edit
    b. MDI
    c. Auto
    d. Jog

10. What does the acronym MDI stand for?

11. Which display screen is used to enter tool information?

12. If the Reset button is pressed during automatic operation, then spindle rotations, feed, and coolant will stop.
    T or F

13. During setup, the mode switch used to allow for manual movement of the machine axes is:
    a. Auto
    b. MDI
    c. Edit
    d. Jog

Unit 3: CNC Turning Center Programming

CNC Turning Center Program Template

Chart 3-3 can be used as a guide for inputting the data necessary to create a program. Make copies to complete each program exercise (or simply use a lined sheet of paper). Certain sections of the turning program can be repetitious in nature. They are the program beginning, the tool beginning, the tool ending and the program ending. The program block structure for each of these sections follow.

Note: If you intend to load any of the programs you create into a machine controller for trial and use, you must include a percent (%) sign on a separate line at the beginning and end of the text. This is required for communications purposes.

The Program Beginning

O7306 = program number
(Comments) = part number or other identifying information
(Comments) = date or other identifying information

Note: 9000 series program numbering is reserved for macro programs; therefore, avoid using it for your program number.

Note: The question marks (T????) in the template sections below represent variable data that are specific to the situation and need to be replaced with live data that is relevant to your programming situation.

N10 G90 G80 G40
N15 G00 G28 U0.0 W0.0
N20 G97 S???? M3
N25 F.???

The Tool Beginning

(Comments) = Tool identification information
N95 G28 U0.0 W0.0
N100 T????
N105 G50 S???
N110 G96 S??? M03
N115 G00 G54 X???? Z???? M08

The Tool Ending

N200 G00 G40 X???? Z.1 M09
N205 G28 U0.0 W0.0 T??00
N210 M01

The Program Ending

N300 G28 U0.0 W0.0 M09
N305 M30

The following G and M-Code reference charts are given to aid in the programming process.

Chart 3-1 Preparatory Functions (G-Codes) Specific To Turning Centers			
Code	**Group**	**Function**	
G00	01	Rapid Traverse Positioning	**NOTES:**
G01	01	Linear Interpolation	
G02	01	Circular and Helical Interpolation CW (clockwise)	*1. In the table, G-Codes marked with an asterisk (*) are Active upon startup of the machine.*
G03	01	Circular and Helical Interpolation CCW (counterclockwise)	
G04	00	Dwell	*2. At machine startup or after pressing reset, the inch (G20) or metric (G21) measuring system last active remains in effect.*
G09	00	Exact Stop	
G10	00	Programmable Data Setting	
G11	00	Programmable Data Setting Cancellation	
G20	06	Input in Inches	*3. G-Codes of group 00 represent "one shot" G-Codes, and they are effective only to the designated blocks.*
G21	06	Input in Millimeters	
*G22	09	Stored Stroke Limit ON	
G23	09	Stored Stroke Limit OFF	*4. Modal G-Codes remain in effect until they are replaced by another command from the same group.*
G25	08	Spindle Speed Fluctuation Detection ON	
G26	08	Spindle Speed Fluctuation Detection OFF	
G27	00	Reference Point Return Check	*5. If modal G-Codes from the same group are specified in the same block, the last one listed is in effect.*
G28	00	Reference Point Return	
G29	00	Return From Reference Point	
G30	00	Return to Second, Third, and Fourth Reference Point	*6. Modal G-Codes of different groups can be specified in the same block.*
G32	01	Thread Cutting	
*G40	07	Tool Nose Radius Compensation Cancel	
G41	07	Tool Nose Radius Compensation, Left Side	*7. If a G-Code from group 01 is specified within a canned drilling cycle block, the cycle will be cancelled just as if a G80 canned cycle cancellation were called.*
G42	07	Tool Nose Radius Compensation, Right Side	
G50	00	Coordinate System Setting/ Maximum Spindle Speed Setting	
G52	00	Local Coordinate System Setting	
G53	00	Machine Coordinate System Setting	*More detailed descriptions and application examples are given later in the section Overview of Preparatory Functions for CNC Turning Centers.*
G54-59	14	Work Coordinate System Selection	
G68	04	Mirror Image for Double Turrets ON	
*G69	04	Mirror Image for Double Turrets OFF	
G70	00	Finishing Cycle	
G71	00	Stock Removal in Turning	
G72	00	Stock Removal in Facing	
G73	00	Pattern Repeating	
G74	00	Peck Drilling Cycle	
G75	00	Groove Cutting Cycle	
G76	00	Multiple Thread Cutting Cycle	
*G80	10	Canned Drilling Cycle Cancellation	
G83	10	Face Drilling Cycle	
G84	10	Face Tapping Cycle	
G86	10	Face Boring Cycle	
G90	01	Outer/Inner Diameter Turning Cycle	
G92	01	Thread Cutting Cycle/Maximum Spindle Speed Setting (SYS B,C)	
G94	01	Face Cutting Cycle	
G96	02	Constant Surface Speed Control	
*G97	02	Constant Surface Speed Control Cancellation	
G98	05	Feed per Minute	
*G99	05	Feed per Revolution	

Chart 3-2 Miscellaneous Functions (M-Codes) Specific To Turning Centers	
M-Code	**Function**
M00	Program Stop
M01	Optional Stop
M02	Program End Without Rewind
M03	Spindle ON Clockwise (CW) Rotation
M04	Spindle ON Counterclockwise (CCW) Rotation
M05	Spindle OFF Rotation Stop
M08	Flood Coolant ON
M09	Coolant OFF
M10	Chuck Close
M11	Chuck Open
M12	Tailstock Quill Advance
M13	Tailstock Quill Retract
M17	Rotation of Tool Turret Forward
M18	Rotation of Tool Turret Backward
M18	Spindle Orient Cancel
M19	Spindle Orient
M21	Tailstock Direction Forward
M22	Tailstock Direction Backward
M23	Threading Finishing with Chamfering
M24	Threading Finishing with Right-Angle
M30	Program End With Rewind
M41	Spindle LOW Gear Range Command
M42	Spindle HIGH Gear Range Command
M71	Bar Feed ON – Start
M72	Bar Feed OFF – Stop
M73	Parts Catcher Advance
M74	Parts Catcher Retract
M76	Parts Counter
M98	Subroutine Call
M99	Return to Main Program From Subroutine

Chart 3-3 provides a programming sheet that can be used as a guide for inputting the data necessary to create a turning center program. Make copies to complete each program exercise or use a separate lined sheet of paper.

Unit 3: CNC Student Workbook

Chart 3-3 CNC Turning Center Programming Sheet

Date:				Prepared By:					
Part Name:				Part Number:					
Machine:				Program Number:					
Line #	Prepatory Code	X-Axis Coordinate	Z-Axis Coordinate	Modifier	Feed Rate	Tool #	Offset #	Spindle Speed	M Codes

CNC Turning Center Tool List

Chart 3-4 identifies the tools to choose from when planning your programs. Note that there is only one tool number listed for a Center Drill and specific drill size. There will undoubtedly be additional locations on the turret for drilling tools. Number the tools accordingly.

Chart 3-4 Cutting Tools for Turning Centers

Tool Graphic	Tool Number	Tool Description
	T0101	#5 High Speed Steel Center Drill
	T0202	O.D. Rough Turning Tool 80 Degree Diamond Insert .031 Tool Nose Radius
	T0303	I.D. Rough Boring Tool 80 Degree Diamond Insert .031 Tool Nose Radius
	T0404	O.D. Finish Turning Tool 55 Degree Diamond Insert .015 Tool Nose Radius
	T0505	I.D. Finish Boring Tool 55 Degree Diamond Insert .01531 Tool Nose Radius
	T0606	O.D. Grooving Tool .005 Tool Nose Radius .118 Wide
	T0707	Standard Drill Sizes Vary as Needed
	T0808	O.D. Threading Tool 60 Degree
Tool Graphics, Courtesy Kennemetal NOVO		

Chart 3-4 Cutting Tools for Turning Centers (continued)

Tool Graphic	Tool Number	Tool Description
	T0909	Inserted Drill Tool .625 Minimum Diameter
	T1010	Alternate O.D. Tool
	T1111	I.D. Groove Tool .005 Tool Nose Radius .118 Wide
	T1212	Part-Off Blade .125 Wide
	Alternate T0909	I.D. Threading 60 Degree Inserted
	Alternate T1111	Reamer Any Required Size
	Alternate T1111	Tap Any Required Size
Tool Graphics, Courtesy Kennemetal NOVO		

Identifying Programming Coordinates for Turning

1. In Figure 3-1, identify and list the programming coordinate points (indicated by the small filled dots) for the part contour using absolute dimensioning. The face of the part (right end) and the centerline are the part zero location. Please list all values for the X-axis as radial values and include the arc center locations for programming of the arcs.

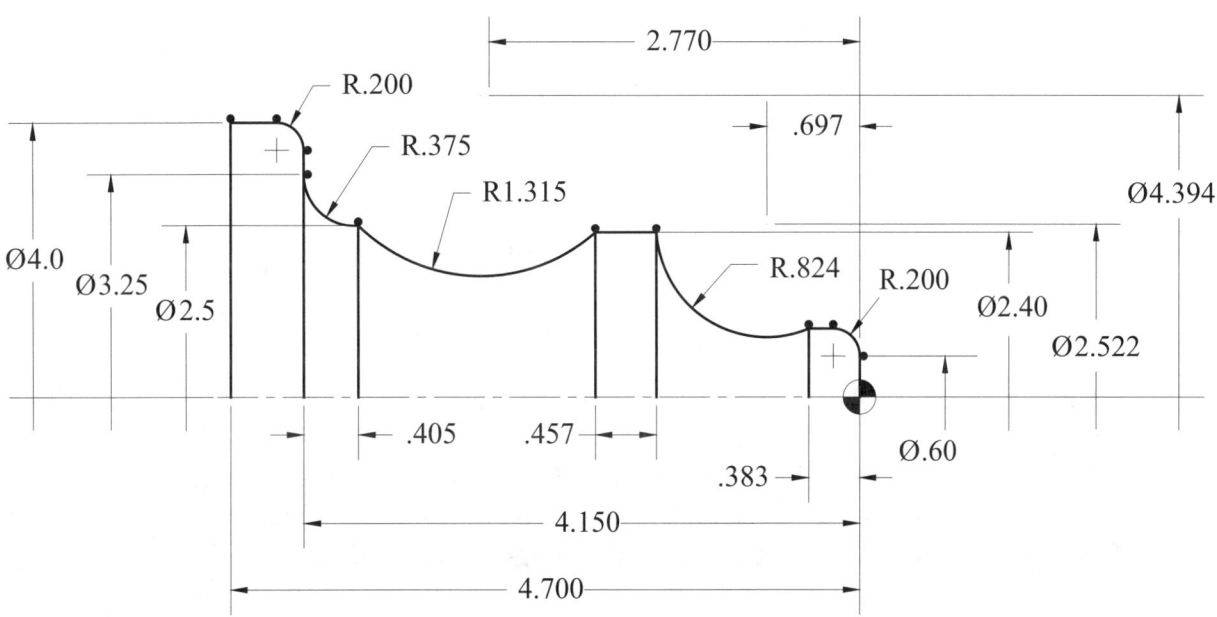

Figure 3-1   Identify the Absolute and Incremental Coordinates

2. Continuing with Figure 3-1, identify and list the programming coordinate points for the part contour using incremental dimensioning. Please list all values for the X-axis as radial values and include the incremental arc center values.

Linear Interpolation

Programming Exercise 3-1

Using linear interpolation, write a program to create the tool path for the contour, in one depth of cut pass, for Figure 3-2. The material is stainless steel. For this exercise and all the remaining exercises, set the r/min to Constant Surface Speed (G96) and use the chart to determine a midrange value. Set the in/rev for feed rate to a midrange value as well. For this exercise (and all remaining examples), you should list the X-axis values as diametrical values. *The dashed line on the drawing indicates the net shape of the part and the metal to be removed.*

Caution: DO NOT attempt to execute this program from solid bar stock.

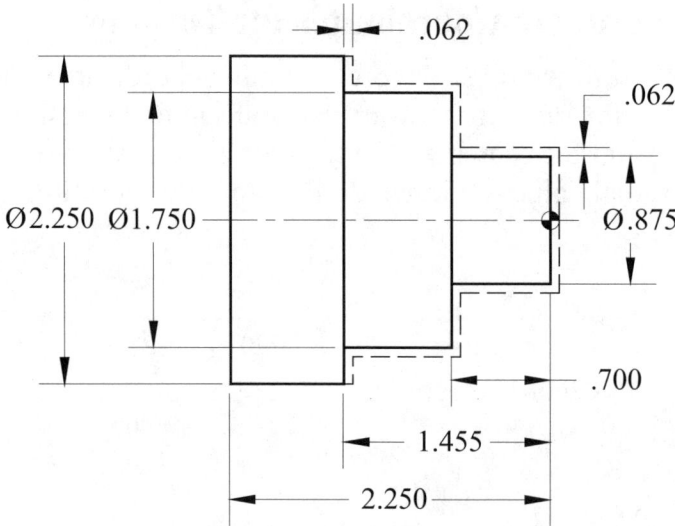

Figure 3-2   Turning Center Programming Exercise 3-1

Programming Exercise 3-2

Using linear interpolation, write a program to create the tool path for the contour, in one depth-of-cut pass, for Figure 3-3. Please use any calculations necessary to offset the tool path for the Tool Nose Radius Compensation (TNRC). The material is alloy steel. *The dashed line on the drawing indicates the net shape of the part and the metal to be removed.*

Caution: DO NOT attempt to execute this program from solid bar stock.

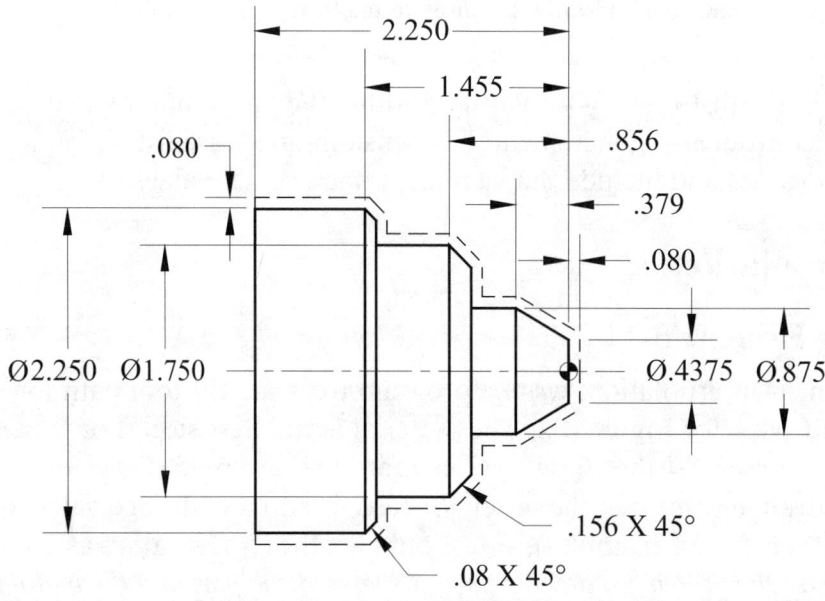

Figure 3-3   Turning Center Programming Exercises 3-2 and 3-7

Programming Exercise 3-3

Using the Fixed Cutting Cycle B (G94), write a program to create the facing cut for Figure 3-4 in three equal depths of cut passes. The material is stainless steel.

Programming Exercise 3-4

Using the Fixed Cutting Cycle A (G90), write a program to create the turning cut for Figure 3-5 in three equal depth of cut passes. The material is stainless steel.

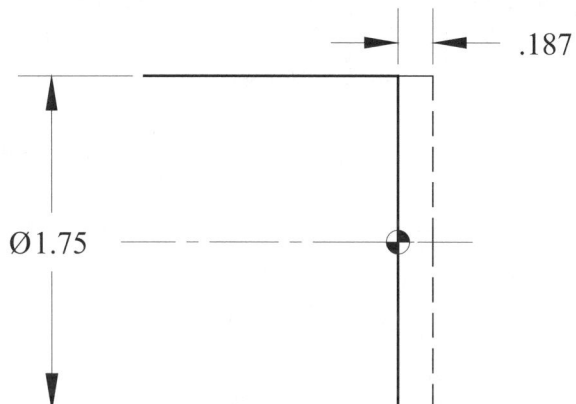

Figure 3-4   Turning Center Programming Exercise 3-3

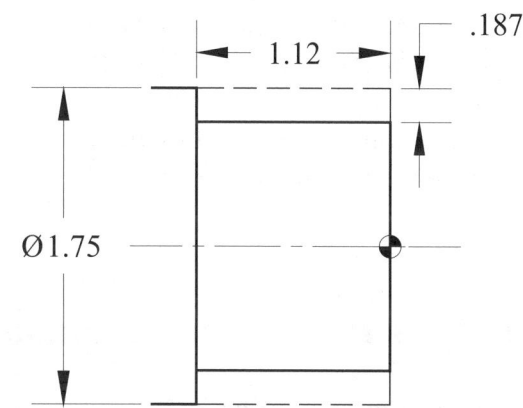

Figure 3-5   Turning Center Programming Exercise 3-4

Linear and Circular Interpolation

Programming Exercise 3-5

Using linear and circular interpolation, write a program to create the tool path for the contour of Figure 3-6. Please use any calculations necessary to offset the tool path to allow for Tool Nose Radius Compensation (TNRC). **Do not** use G41 or G42 in this exercise. The material is aluminum. The machine being used has a maximum spindle r/min of 6000. *The dashed line on the drawing indicates the net shape of the part and the metal to be removed.*

Caution: DO NOT attempt to execute this program from solid bar stock.

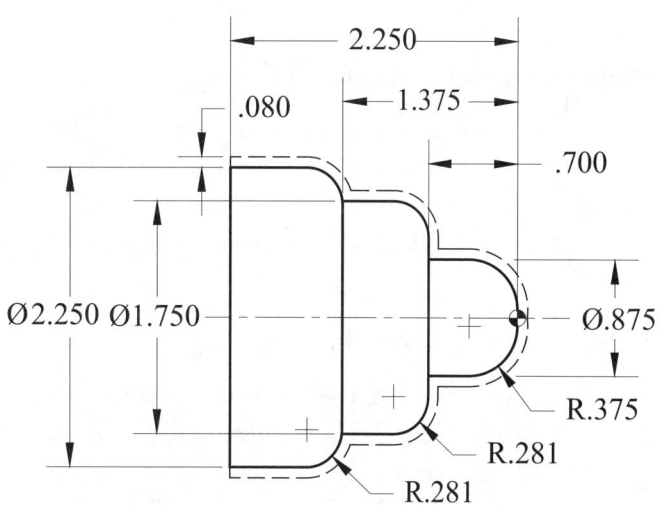

**Figure 3-6   Turning Center Programming
Exercises 3-5 and 3-8**

Programming Exercise 3-6

Using linear and circular interpolation, write a program to create the tool path for the contour of Figure 3-7. Please use any calculations necessary to offset the tool path for Tool Nose Radius Compensation (TNRC). **Do not** use G41 or G42 in this exercise. The material is aluminum. *The dashed line on the drawing indicates the net shape of the part and the metal to be removed.*

Caution: DO NOT attempt to execute this program from solid bar stock.

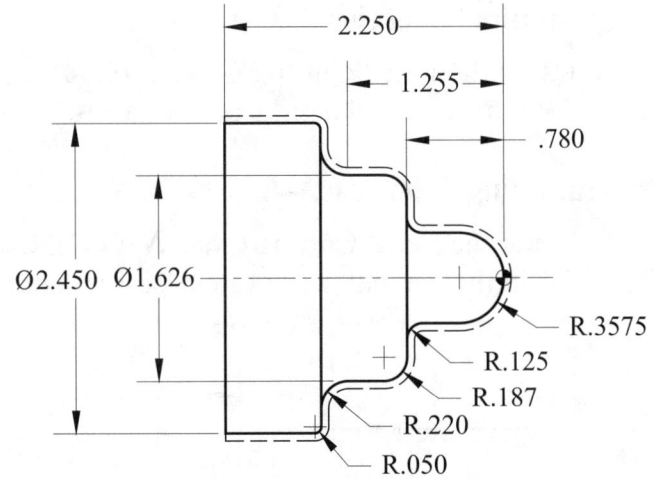

Figure 3-7   Turning Center Programming Exercise 3-6 and 3-9

Tool Nose Radius Compensation

Programming Exercise 3-7

Now use the appropriate Tool Nose Radius Compensation (TNRC) G40, G41, and/or G42 for the tool path contours in Programming Exercise 3-2 (Figure 3-3). (For the remainder of all the excercises, you should use the appropriate TNRC.) *The dashed line on the drawing indicates the net shape of the part and the metal to be removed.*

Caution: DO NOT attempt to execute this program from solid bar stock.

Programming Exercise 3-8

In this exercise, use the appropriate TNRC for the tool path contours in Programming Exercise 3-5 (Figure 3-6). *The dashed line on the drawing indicates the net shape of the part and the metal to be removed.*

Caution: DO NOT attempt to execute this program from solid bar stock.

Programming Exercise 3-9

In this exercise, use the appropriate TNRC for the tool path contours in Programming Exercise 3-6 (Figure 3-7). *The dashed line on the drawing indicates the net shape of the part and the metal to be removed.*

Caution: DO NOT attempt to execute this program from solid bar stock.

Drilling

Programming Exercise 3-10

In this example, it is necessary to center drill the part in order to prepare for subsequent drilling (Figure 3-8). The material is stainless steel and the finished hole should be countersunk to a .405 inch diameter. Calculate the required depth for the center drill and program the tool path using G01.

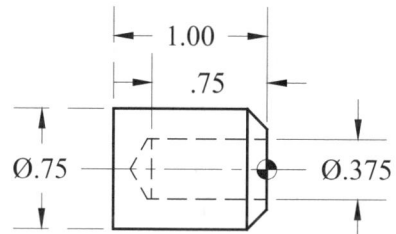

Figure 3-8   Turning Center Programming Exercise 3-10

Programming Exercise 3-11

G74 Drilling Cycle

To finish the hole as described in Exercise 3-10, it is necessary to drill to the depth as shown in Figure 3-9. The material is stainless steel and the G74 drilling cycle should be used with three equal depth cuts.

Figure 3-9   Turning Center Programming Exercise 3-11

Multiple Repetitive Cycles

Programming Exercise 3-12

G71 and G70 Rough and Finish Turn Cycle

The part in Figure 3-10 needs to be rough and finish turned to specifications. Use the G71 and G70 Turning Cycles to accomplish this. In this case, the material is aluminum. Set the roughing depth of cut at .08 inch and the finish allowance for the X-axis at .015 inch and .005 inch on the Z-axis. The cut on the face of the part of .031 of excess material should be made in one pass. Make any calculations necessary to program the tool path and write the program.

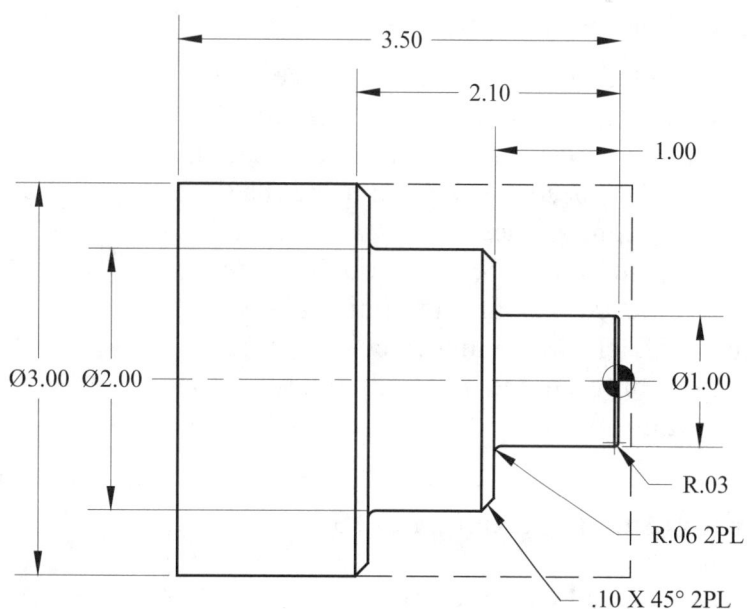

Figure 3-10   Turning Center Programming Exercise 3-12

Boring

Programming Exercise 3-13

G71 and G70 Rough and Finish Turn Cycle

The same Rough and Finish Turning Cycles may be applied to internal boring. Write a program for Figure 3-11 using these cycles. The material is aluminum and the .813 inch diameter hole already exists in the part. Set the rough depth of cut at .08 inch and the finish allowance at .015 inch and .005 inch on the Z-axis. Make any calculations necessary to program the tool path and write the program, including any preparatory machining necessary.

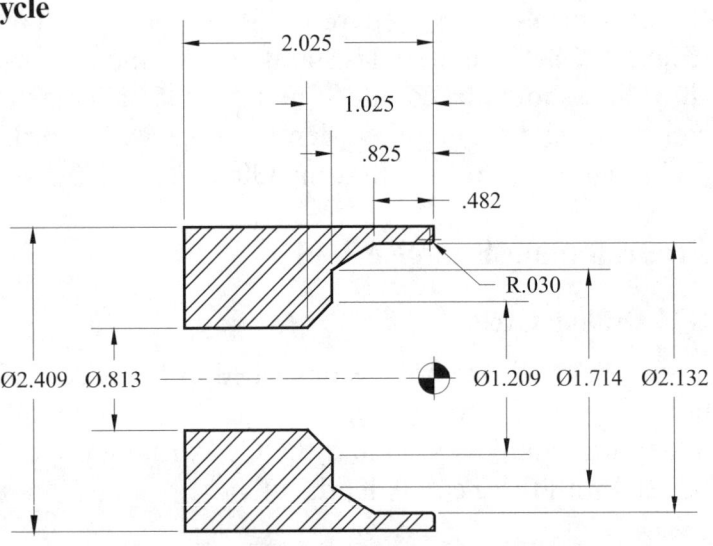

Figure 3-11   Turning Center Programming Exercise 3-13

Programming Exercise 3-14

G72 Face Cutting Cycle

Occasionally, the part geometry will dictate the method of machining. The following case is one instance. The part in Figure 3-12 needs to be faced using the G72 Face Cutting Cycle. In this case, the material is aluminum. Set the roughing depth of cut at .06 inch and the finish allowance at .015 inch. Make any calculations necessary to program the tool path and write the program.

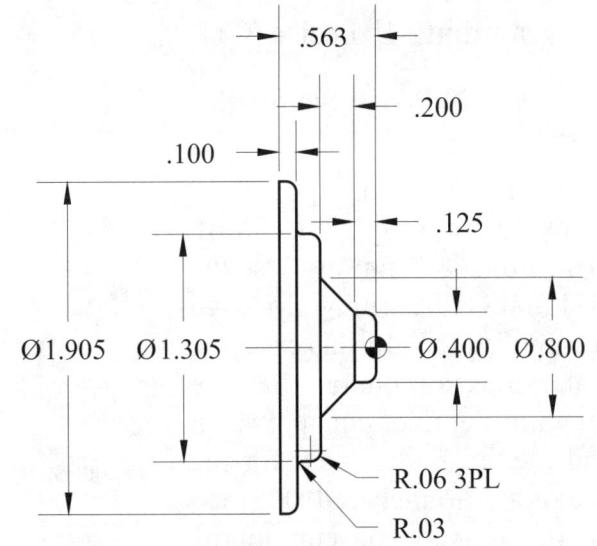

Figure 3-12   Turning Center Programming Exercise 3-14

Programming Exercise 3-15

G73 Pattern Repeating Cycle

The part in Figure 3-13 has a net shape that needs a specific amount of material removed from all surfaces. Use the G73 turning cycle to accomplish this. In this case, the material is stainless steel. Set the number of roughing passes at three and the finish allowance at .03 inch. Make any calculations necessary to program the tool path and write the program. *The dashed line on the drawing indicates the net shape of the part and the metal to be removed.*

Caution: DO NOT attempt to execute this program from solid bar stock.

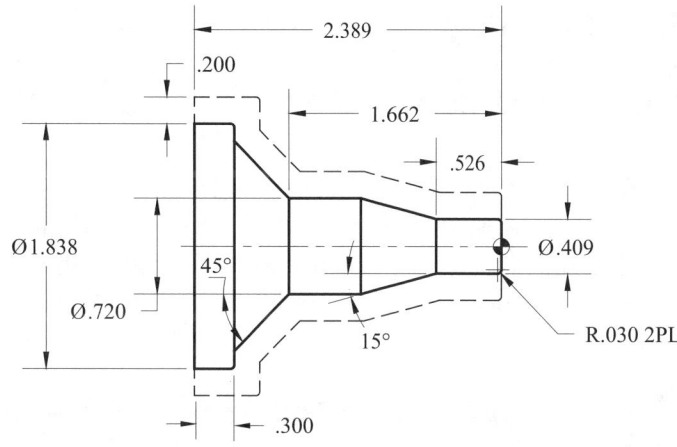

Figure 3-13 Turning Center Programming Exercise 3-15

Programming Exercise 3-16

The same situation applies to a case of internal boring. Write a program for Figure 3-14 using the G73 cycle. The material is stainless steel. Set the number of roughing passes at three and the finish allowance at .03 inch. Make any calculations necessary to program the tool path and write the program. *The dashed line on the drawing indicates the net shape of the part and the metal to be removed.* The .813 diameter has been predrilled in an earlier operation.

Caution: DO NOT attempt to execute this program from solid bar stock.

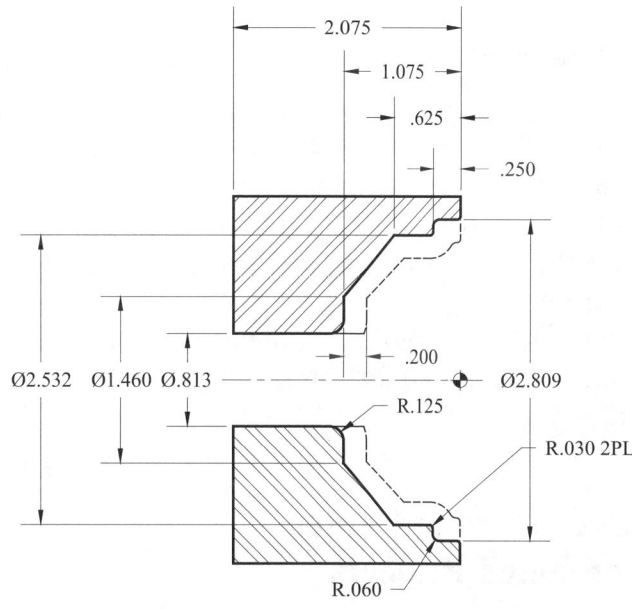

Figure 3-14 Turning Center Programming Exercise 3-16

Grooving

Programming Exercise 3-17

G75 Grooving Cycle

A common operation prior to threading is the creation of an escape groove at the end of the threads. Use the information in the following drawing to write a program to machine the groove using the G75 cycle. The material is carbon steel.

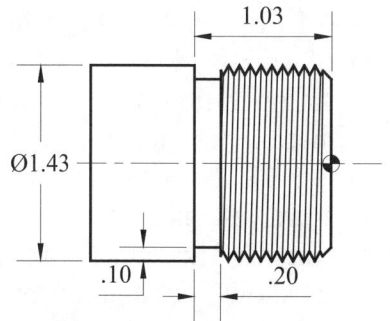

Figure 3-15 Turning Center Programming Exercise 3-17

OD Threading

Programming Exercise 3-18

G76 Threading Cycle

Write a program to machine the threaded portion on the part in Figure 3-16, using the G76 threading cycle. The material is carbon steel bar stock of 1.0 inch diameter.

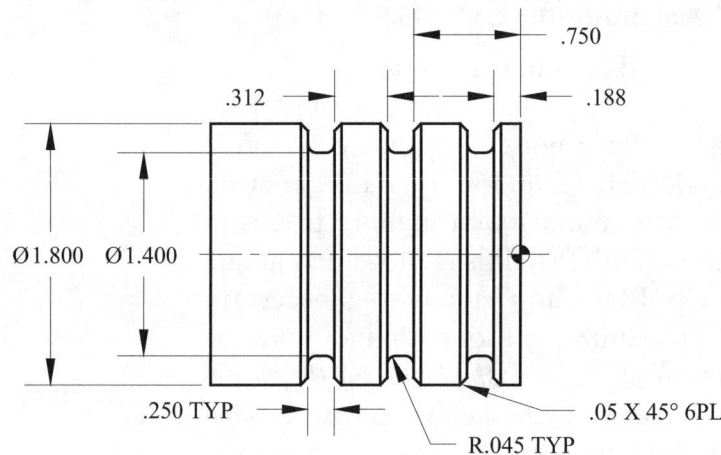

1.0 - 8 UNC 2A

Figure 3-16   Turning Center Programming Exercise 3-18

CNC Turning Center Subprogram Application

Programming Exercise 3-19

M98 and M99

Use linear and circular interpolation to create the programmed tool path for the grooves shown in Figure 3-17. You should make the groove geometry in a subprogram and then call it for each groove location from the main program. The material for this project is aluminum.

Figure 3-17   Turning Center Programming Exercise 3-19

CNC Turning Center Combined Projects

Programming Exercise 3-20

In the Process Planning section of this workbook, you identified the operation, tools, and setup information for the part shown in Figure 3-18. Now use the information that you gathered to write a program for the same part. Note: the OD should be roughed and then finished with separate tools.

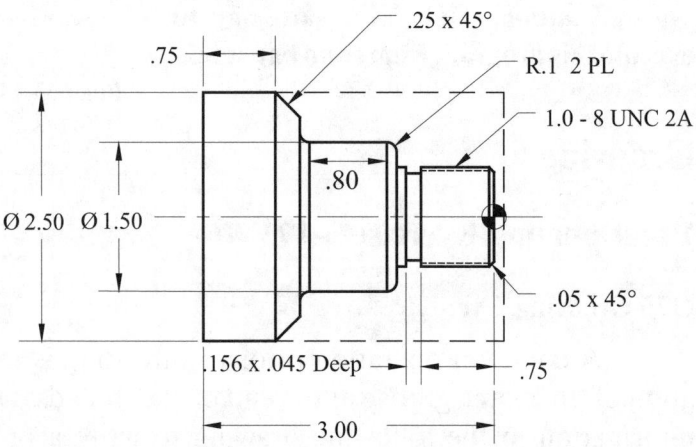

Figure 3-18   Turning Center Programming Exercise 3-20

CNC Turning Center Program Error Diagnosis

Use the skills you have learned to identify the problems in the following program lines and program sections. You may refer to the text, *Programming of CNC Machines*, Fourth Edition.

1. Use the following CNC code and sketch a representation of the part that would be machined.

```
O2001
(CNC Turning Center Program Diagnosis, Problem 1)
(Tool #1, Rough Turning Tool)
N10 G28 U0.0 W0.0
N15 T0100 M42
N20 G50 S2000
N25 G96 S500 M03
N30 G00 G54 X2.2 Z.3 T0101 M08
N35 G01 Z.01 F.03
N40 X0.0 F.012
N45 G00 X3.0 Z.2
N50 G73 P50 Q85 I.168 K.169 U.04 W.02 D3 F.012
N55 G00 X1.59
N60 G01 Z0.0
N65 X1.75 Z–.08
N70 Z–1.375
N75 X2.0 W–.125
N80 Z–2.1
N85 G03 U.3 Z–2.25 I–.15 K0.0 F.004
N90 G01 X2.85
N95 G00 G40 Z.1 M09
N100 G28 U0.0 W0.0 T0100
N105 M30
```

2. Find the error in this program line.

```
N100 G03 U.3 Z–2.25 F.004
```

3. Find the error in the following program lines.

   O2003
   (CNC Turning Center Program Diagnosis, Problem 3)
   (Tool 9 = #5 Center Drill)
   N10 T0900
   N20 G97 S641 M03
   N30 G00 X0Z.1 T0909 M08
   N40 G1 Z–.4303
   N50 G0 Z.1 M09
   N60 G28 U0.0 W0.0 T0900
   N70 M30

4. Find the error in the following program section.

   O2005
   (CNC Turning Center Program Diagnosis, Problem 5)
   (Tool #1, Rough Turning Tool)
   N10 T0100
   N20 G50 S6000
   N30 G96 S3650
   N40 G00 X0 Z.1 T0101 M08
   . . . .
   . . . .

5. Find the error in the following program.

   O2005
   (CNC Turning Center Program Diagnosis, Problem 5)
   (Tool #3, Rough Boring Tool)
   N10 T0300
   N20 G50 S6000
   N30 G96 S2800 M3
   N40 G0 G41 X.713 Z.1 T0303 M08
   N50 G71 U.08 R.03
   N60 G71 P50 Q150 U.015 W.005
   N70 G0 X2.192
   N80 G01 Z0 F.0265
   N90 G2 X2.132 Z–.03 R.03
   N100 G01 Z–.482
   N110 X1.714 Z–.825
   N120 X1.209
   N130 X.813 Z–1.025
   N140 X.713
   N150 G28 G40 U0.0 W0.0 T0300
   N160 M01

6. Find the error in the following program section.

O2007
(CNC Turning Center Program Diagnosis, Problem 6)
(Tool #1, Rough Turning Tool)
N10 T0100
N20 G96 S563 M03
N30 G00 G41 X.6975 Z.1 M08
N40 G1 Z0.0 F.022
N50 X–.01
N60 G0Z.1
N70 G42 X.4375
. . . .
. . . .

7. Find the error in the following program.

O2008
(CNC Turning Center Program Diagnosis, Problem 7)
(Tool #1, Rough Turning Tool)
N10 T0100
N20 G96 S563 M03
N30 G00 G41 X.6975 Z.1 T0101 M08
N40 G1 Z0.0 F.022
N50 X–.01
N60 G0Z.1
N70 X.4375
N80 G1Z0
N90 X.875 Z–.379
N100 Z–.7
N110 X1.438
N120 X1.75 Z–.856
N130 Z–1.375
N140 X2.09
N150 X2.25 Z–1.455
N160 Z–2.25
N170 G0 X2.35 Z.1 M09
N180 G28 G40 U0.0 W0.0
N190 M30

8. Find the error in the following program section.

 O2009
 (CNC Turning Center Program Diagnosis, Problem 8)
 (Tool #1, Rough Turning Tool)
 N10 T0100
 N20 G50 S6000
 N30 G96 S2800 M3
 N40 G0 G42 X2.005 Z.1 T0101 M08
 N50 G72 U.06 R.1
 N60 G72 P70 Q190 U.015 W.015
 N70 G0 Z0 F.0265
 N80 X.28
 N90 G3 X.4 Z–.06 R.06
 N100 Z–.125 F.0265
 N110 X.8 Z–.325
 N120 X1.185
 N130 G3 X1.305 Z–.385 R.06
 N140 G01 Z–.433
 N150 G2 X1.365 Z–.463 R.03
 N160 G1 X1.785
 N170 G3 X1.905 Z–.523 R.06
 N180 G01 Z–.563
 N190 X2.005
 N200 G28 G40 U0.0 W0.0 T0100
 N210 M01

9. Find the error in the following program section.

 (Tool #2, Finish Turning Tool)
 N220 T0200
 N230 G50 S6000
 N240 G96 M3
 N250 G0 G42 X.90 Z.1 T0202 M08
 N260 G70 P80 Q190
 N270 G28 G40 U0.0 W0.0 T0200 M09
 N280 M30

10. Find the error in the following program section.

    (Tool #4, Finish Boring Tool)
    N170 T0400
    N180 G50 S6000
    N190 G96 S2800 M3
    N200 G0 G42 X.90 Z.1 T0404 M08
    N210 G70 P80 Q140
    N220 G28 U0.0 W0.0 T0400 M09
    N230 M30

Unit 4: CNC Machining Center Programming

CNC Machining Center Programming Template

Certain sections of the machining program can be repetitious in nature. They are the program beginning, the safety block, the tool beginning, the tool ending, and the program ending. *Note: If you intend to load any of the programs you create into a machine controller for trial and use, you must include a percent (%) sign on a separate line at the beginning and end of the text. This is required for communications purposes.*

The Program Beginning

O2406 = program number
(Comments) = part number or other identifying information
(Comments) = date or other identifying information
    Note: 9000 series program numbering is reserved for Macro programs; therefore, avoid using it for your program number

N10 G90 G20 G80 G40 G49

The Safety Block

For an explanation of the safety block, see *Programming of CNC Machines*, Fourth Edition, Part 4 (Programming of CNC Machining Centers), for complete details.

N15 G28 X0.0 Y0.0 Z0.0

The Tool Beginning

(Comments) = Tool identification information
N20 T01 M06
N25 S1000 M03
N30 G54 G00 X0.0Y0.0
N35 G43 Z1.0 H01
N40 Z.1 M08
    Note: The tool number (T01) and tool height offset number (H01), the value entered in your program for spindle r/min (S), and the values entered in your programs for X and Y coordinates listed with G54 will vary dependent on the specific application.

The Tool Ending

N100 G80 Z.1 M09
N105 G91 G28 Z0.0
N110 M01

The Program Ending

N200 G28 X0.0 Y0.0
N205 M30

    Note: the coordinate points in line N30 of the above template will be replaced with live data that is relevant to your programming situation.

    Reference Charts 4-1 (G-Codes) and 4-2 (M-Codes) are provided to aid in the programming process.

Chart 4-1 Preparatory Functions (G-Codes) Specific To Machining Centers

Code	Group	Function	Code	Group	Function
*G00	01	Rapid Traverse Positioning	G63	15	Tapping Mode
*G01	01	Linear Interpolation	G64	15	Cutting Mode
G02	01	Circular and Helical Interpolation CW (clockwise)	G68	16	Rotation of Coordinate System
G03	01	Circular and Helical Interpolation CCW (counterclockwise)	*G69	16	Cancellation of Corrdinate System Rotation
G04	00	Dwell	G73	09	Peck Drilling Cycle
G09	00	Exact Stop	G74	09	Reverse Tapping Cycle
G10	00	Programmable Data Setting	G76	09	Fine Boring Cycle
G11	00	Programmable Data Setting Cancellation	*G80	09	Canned Cycle Cancellation
*G15	17	Polar Coordinate Cancelation	G81	09	Drilling Cycle, Spot Drilling
G16	17	Polar Coordinate System	G82	09	Drilling Cycle, Counter Boring
*G17	02	XY Plane Selection	G83	09	Deep Hole Peck Drilling Cycle
G18	02	ZX Plane Selection	G84	09	Tapping Cycle
G19	02	YZ Plane Selection	G85	09	Reaming Cycle
G20	06	Input in Inches	G86	09	Boring Cycle
G21	06	Input in Millimeters	G87	09	Back Boring Cycle
*G22	04	Stored Stroke Limit ON	G88	09	Boring Cycle
G23	04	Stored Stroke Limit OFF	G89	09	Boring Cycle
G27	00	Reference Point Return Check	*G90	03	Absolute Programming
G28	00	Reference Point Return	*G91	03	Incremental Programming
G29	00	Return From Reference Point	G92	00	Setting for the Work Coordinate System or Maximum Spindle r/min
G30	00	Return to Second, Third, and Fourth Reference Point	*G94	05	Feed per Minute
G33	01	Thread Cutting	G95	05	Feed per Revolution
G37	00	Automatic Tool Length Measurement	G96	02	Constant Surface Speed Control
*G40	07	Cutter Compensation Cancel	*G97	02	Constant Surface Speed Control Cancel
G41	07	Cutter Compensation, Left	*G98	10	Canned Cycle Initial Level Return
G42	07	Cutter Compensation, Left	G99	10	Canned Cycle R=Level Return
G43	08	Tool Length Offset Compensation positive (+) direction			
G44	08	Tool Length Offset Compensation negative (−) direction			**NOTES:**
G45	00	Tool Offset Increase			
G46	00	Tool Offset Decrease			*The items marked with an asterisk (*) are active upon startup of the machine or are reinstated when the RESET button has been pressed. Check the specific manufacturer Operation Manual for your application.*
G47	00	Tool Offset Double Increase			
G48	00	Tool Offset Double Decrease			
G49	08	Tool Length Offset Compensation Cancel			*For G00, G01, G90 and G91 the initial code that is active is determined by a parameter setting. These are typically G01 and G90 for startup condition.*
*G50	11	Scaling Cancel			
G51	11	Scaling			*G-Codes from groups 00 are one-shot G-Codes.*
G52	00	Local Coordinate System Setting			
G53	00	Machine Coordinate System Setting			*Multiple G-Codes from different groups can be specified in the same block. If more thatn one from the same group is specified only the last G-Code listed will be active.*
*G54-59	14	Work Coordinate System Selection (G54 default)			
G60	00	Single Direction Positioning			

Chart 4-2 Miscellaneous Functions (M-Codes) Specific To Machining Centers

M-Code	Function
M00	Program Stop
M01	Optional Stop
M02	Program End Without Rewind
M03	Spindle ON Clockwise (CW) Rotation
M04	Spindle ON Counterclockwise (CCW) Rotation
M05	Spindle OFF Rotation Stop
M06	Tool Change
M07	Mist Coolant ON
M08	Flood Coolant ON
M09	Coolant OFF
M10	Work Table Rotation Locked
M11	Work Table Rotation Unlocked
M13	Spindle ON CW with Coolant
M14	Spindle ON CCW with Coolant
M16	Change of Heavy Tools
M19	Spindle Orientation
M21	Mirror Image, X-axis
M22	Mirror Image, Y-axis
M23	Mirror Image Cancelaltion
M30	Program End With Rewind
M98	Subroutine Call
M99	Return to Main Program From Subroutine

Chart 4-3 is a CNC Machining Center Programming Sheet that can be used as a guide for inputting the data necessary to create a program. Make as many copies as are needed to complete each program exercise, or use a separate lined sheet of paper.

Chart 4-3   CNC Machining Center Programming Sheet

Date:					Prepared By:					
Part Name:					Part Number:					
Machine:					Program Number:					
Line #	Prepatory Code	X-Axis Coordinate	Y-Axis Coordinate	Z-Axis Coordinate	Modifier	Feed Rate	Tool #	Offset #	Spindle Speed	M Codes

CNC Machining Center Tool List

The following list identifies the tools that are available for planning your programs. For these exercises, the tool carousel on the CNC Machining Center you are programming holds up to 30 tools and the diameter offset value will correspond with the tool number (i.e. T1 = D1). In the following exercises, where only one tool is required, a CNC Setup sheet is not necessary. However, the tool and setup information should be listed in a comment before your program code. Please use a CNC Setup sheet in all other cases, for the sake of clarity.

Face Mill, 3.0 inch diameter, 90°, 5 teeth, Carbide

End Mill, 2-Flute, 1/8 inch
End Mill, 2-Flute, 3/8 inch
End Mill, 2-Flute, 9/16 inch
End Mill, 2-Flute, 1.0 inch

End Mill, 4-Flute, 3/8 inch
End Mill, 4-Flute, 1/2 inch
End Mill, 4-Flute, 5/8 inch
End Mill, 4-Flute, 3/4 inch
End Mill, 4-Flute, 1.0 inch

Roughing End Mill, 4-Flute, 1.0 inch

#5 Center Drill
#6 Center Drill
Spotting Drill, .75 diameter, 90° single flute
Drill Bits, All sets are available in High Speed Steel
Taps, All sizes are available in High Speed Steel
Reamers, all required sizes are available in High Speed Steel

Identifying Programming Coordinates for Milling

1. Use Figure 4-1 to identify the points on the profile geometry using absolute dimensioning to program the part. Include the arc center locations. Start at the zero location of the part and proceed clockwise until all points are identified.

2. Use Figure 4-1 to identify the points on the profile geometry using incremental dimensioning to program the part. Include the arc center locations. Start at the zero location of the part and proceed clockwise until all points are identified.

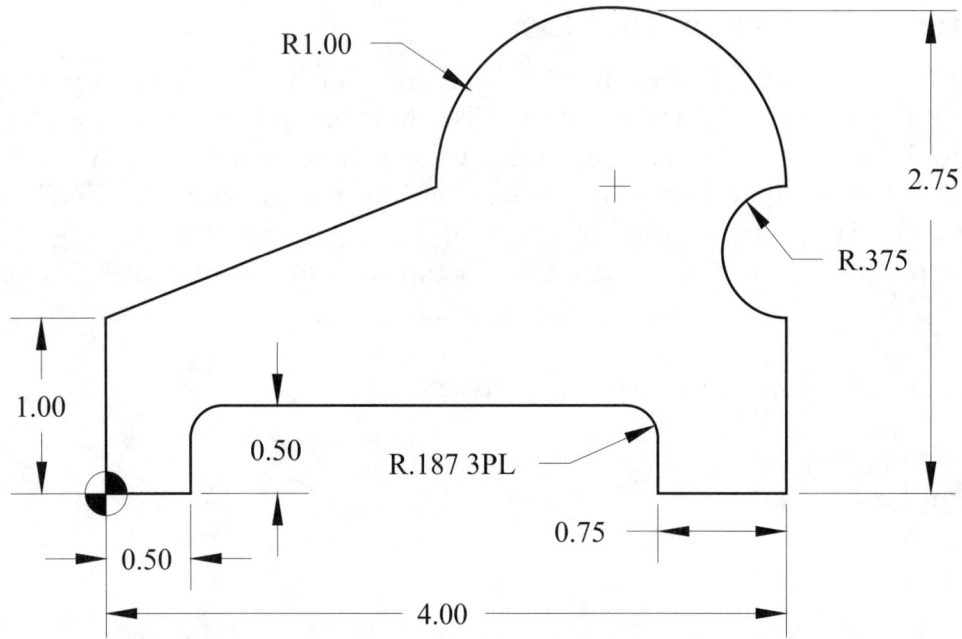

Figure 4-1   Identify Absolute and Incremental Coordinates

Linear Interpolation on the CNC Machining Center

In each of the following examples, if only one tool is needed to complete the part, a CNC Setup Sheet is not required. Please list the tool to be used, the Cutting Speed range, the r/min range, the in/tooth range, and the in/min range, where appropriate. Choose the mid-range values for feeds and speeds to input into your programs.

Face Milling

Programming Exercise 4-1

Using G00 and G01, program the moves required to face off .08 inch from the face of the part in Figure 4-2. The material is alloy steel.

**Figure 4-2   Machining Center
Programming Exercise 4-1**

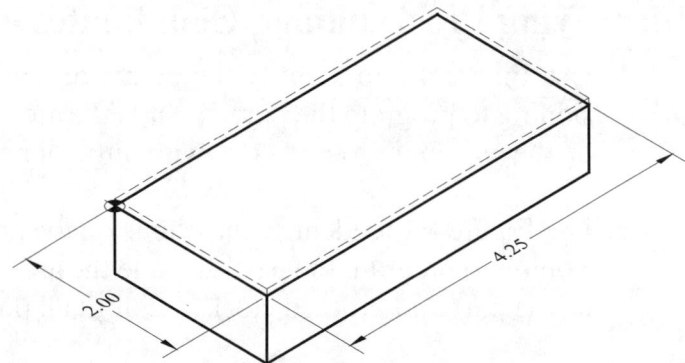

Programming Exercise 4-2

Using G00 and G01, program the moves required to face off .07 inch from the face of the part in Figure 4-3. Use single directional cutting. The material is aluminum.

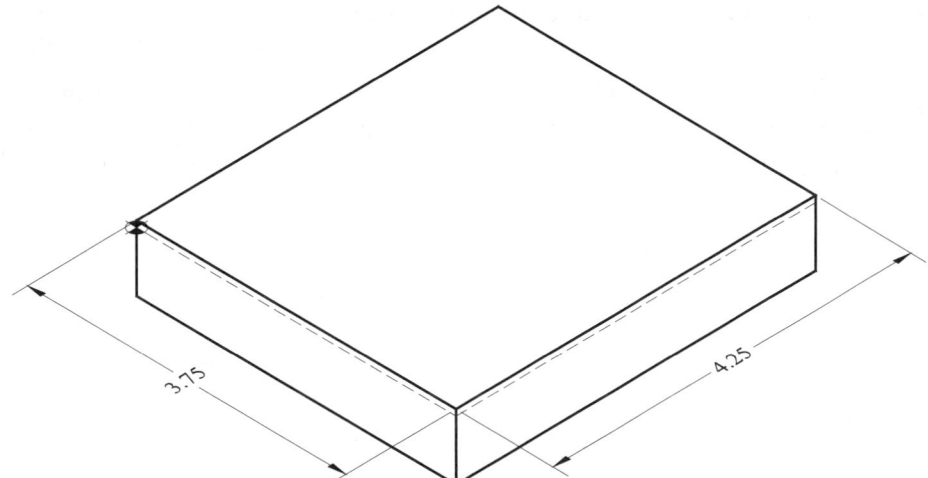

Figure 4-3   Machining Center Programming Exercise 4-2

Contour Milling

Programming Exercise 4-3

Using G00 and G01, program the moves required to mill the contour of the part in Figure 4-4. Use a .375 inch diameter 4 fluted (HSS) end mill and climb mill directional cutting for this example. The material is alloy steel and the part zero is located in the upper left corner of the part. Machining the step will require two depth-of-cut passes.

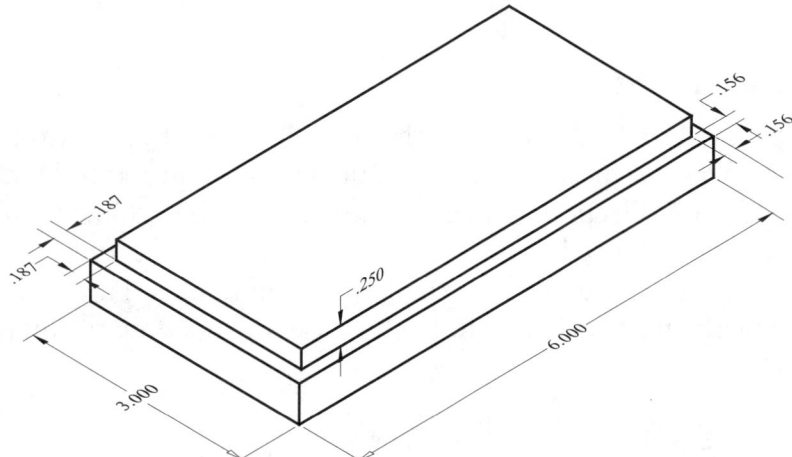

Figure 4-4   Machining Center Programming Exercise 4-3

Circular Interpolation

Programming Exercise 4-4

Using G01, G02, or G03, program the moves required to mill the contour of the part in Figure 4-5. To enable cutting of the entire step in one axial pass, you will need to make a calculation to select the correct minimum diameter end mill. Use climb mill directional cutting for this example. For a minimum of two out of four of the radii, you must use the I and J commands for programming. The material is alloy steel. The depth of the step is .375 inch and it should be machined in two equal depth-of-cut passes. The overall part thickness is 1.0 inch.

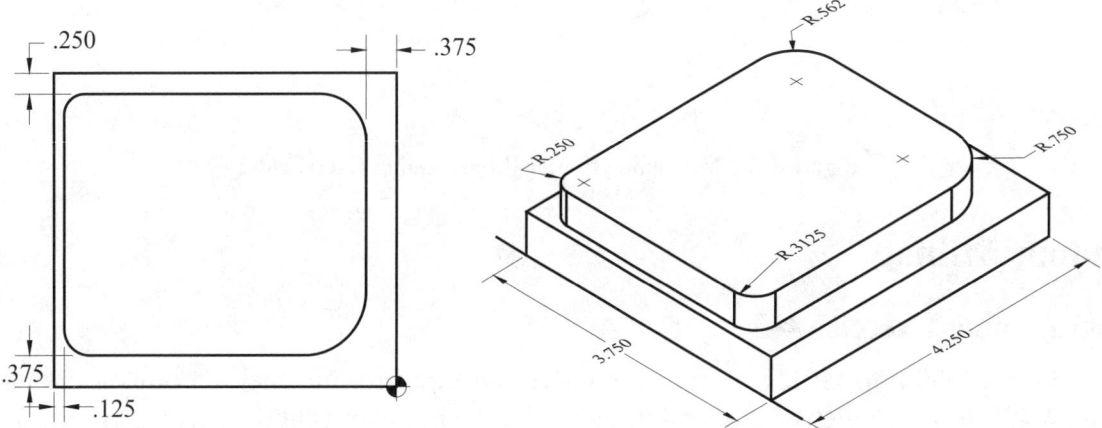

Figure 4-5   Machining Center Programming Exercise 4-4

Programming Exercise 4-5

Using G01, G02, or G03, program the moves required to mill the contour of the part in Figure 4-6. You will need to make a calculation to properly offset the cutter for the angular cut. End mill selection will be dictated by the fillet radius requirements on the profile of the part. You will need to use climb mill directional cutting. For a minimum of two out of four of the radii, you must use the I and J commands. The material is aluminum and the thickness of the part is .500 inch. In this case, the maximum r/min of the machine being used is 6000.

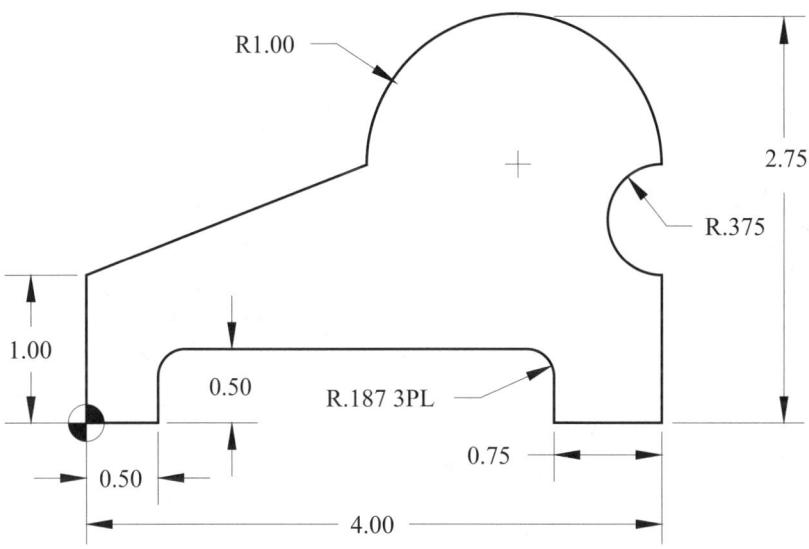

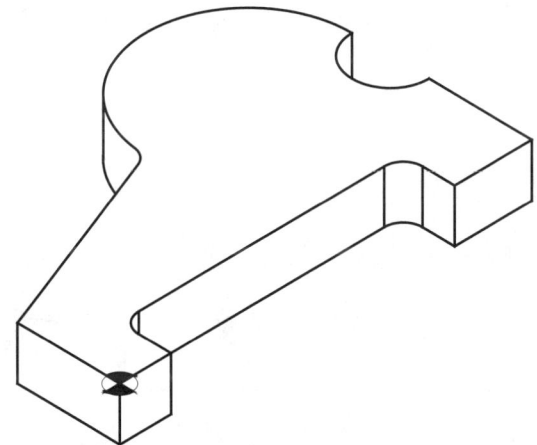

Figure 4-6   Machining Center Programming Exercise 4-5

Circle Milling

Programming Exercise 4-6

Using G01, G02, or G03, program the moves required to mill the circular contour of the part in Figure 4-7. To enable cutting the entire step in one axial pass, you will need to make a calculation to select the correct minimum diameter end mill. Use climb mill directional cutting. The material is carbon steel and a hole of .500 inch diameter exists through the center of the part prior to this machining operation. The step depth is .375 inch and the circular pocket depth is .500 inch.

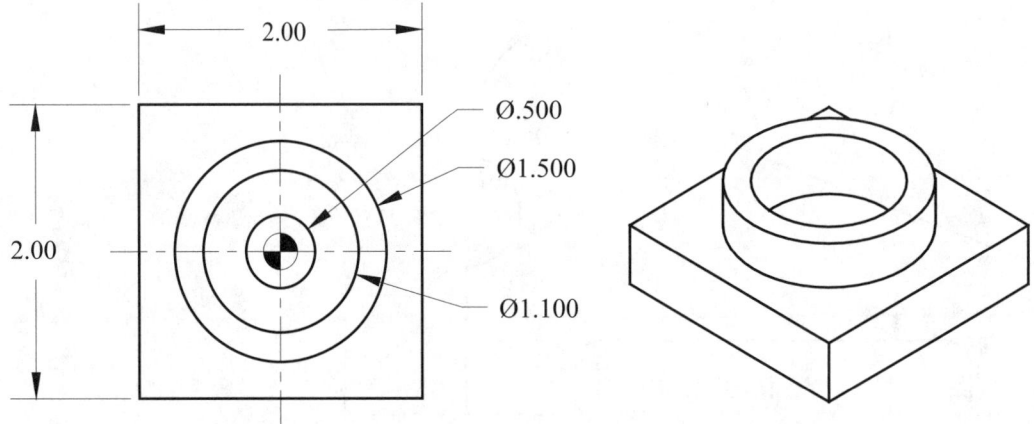

Figure 4-7   Machining Center Programming Exercise 4-6

Cutter Diameter Compensation

Using G01, G02, or G03 along with G41, G42, and G40 for cutter diameter compensation, follow the original directions and program the tool path moves required for each of the last four contour milling exercises.

Programming Exercise 4-7

Program the tool path for Exercise 4-3.

Programming Exercise 4-8

Program the tool path for Exercise 4-4.

Programming Exercise 4-9

Program the tool path for Exercise 4-5.

Programming Exercise 4-10

Program the tool path for Exercise 4-6.

Canned Cycles

Programming Exercise 4-11

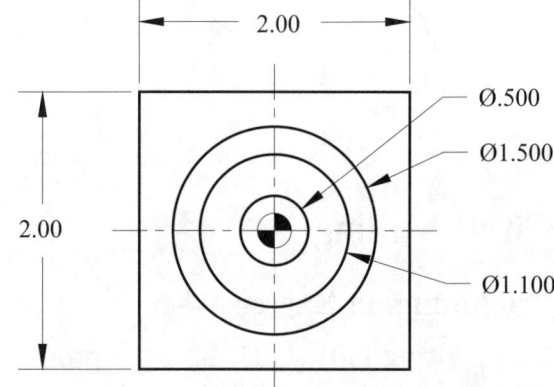

**Figure 4-8   Machining Center Programming
Exercise 4-11**

G81 Drilling

In this exercise, you are required to center drill for and then drill a .500 inch hole at the origin of the part to 1/2-inch depth to the drill point (Figure 4-8). This activity should be added to the beginning of the prior Circle Milling exercise, before circle milling can begin, to aid in the entry of the end mill.

Programming Exercise 4-12

G81 and G73 Drilling

In Programming Exercise 4-3, you programmed the milled step around the contour. In this exercise, you should add the holes to the existing program where required (Figure 4-9). They should be spot drilled deep enough to chamfer the top of the holes .015 inch per side. Each of the holes are to be drilled through the material thickness of .750 inch.

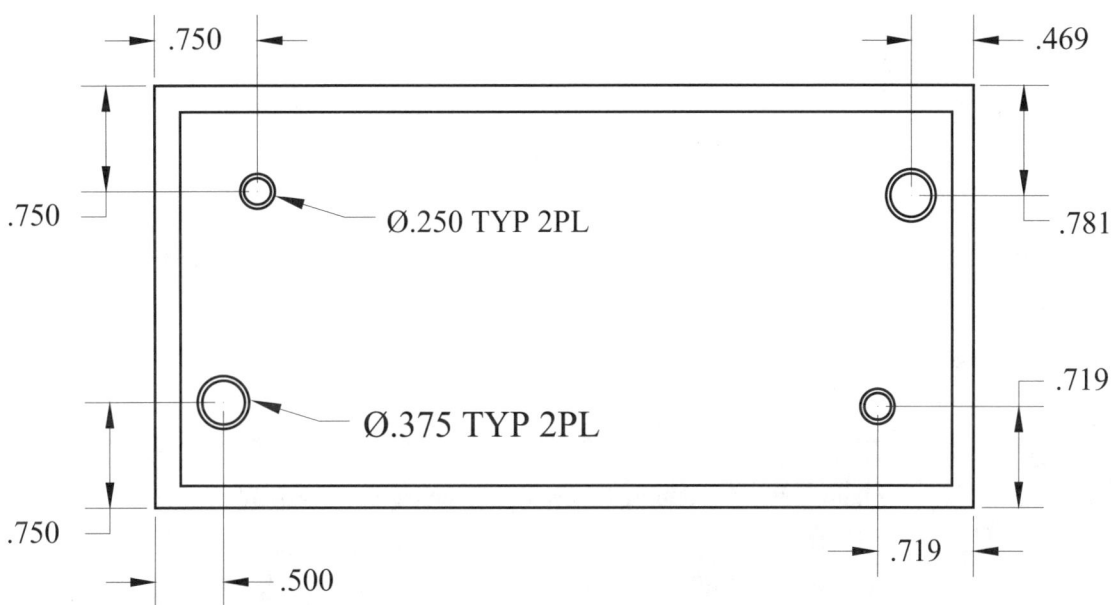

Figure 4-9   Machining Center Programming Exercise 4-12

Programming Exercise 4-13

G81, G83, and G82 Drilling

In Programming Exercise 4-4, you programmed the tool path for the milled step. In this example, you should add the counter-bored holes to the existing program at .375 inch depth (Figure 4-10).

Programming Exercise 4-14

G81, G83 Drilling, and G84 Tapping

In the part in Figure 4-11, it is necessary to machine all of the holes in a plate that is 1.0 inch thick and is made of aluminum. Use the G81, G82, G83, and G84 Canned Cycles to make the program. The part is already machined to the overall size and thickness dimensions. The maximum spindle r/min of the machine is 6000.

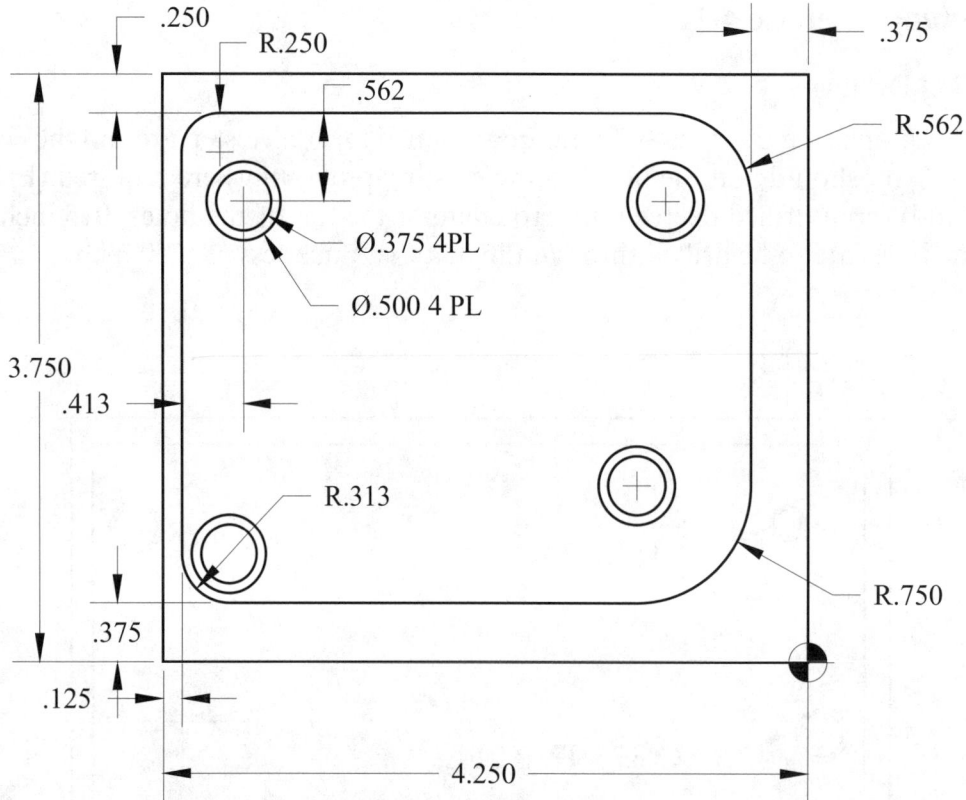

Figure 4-10   Machining Center Programming Exercise 4-13

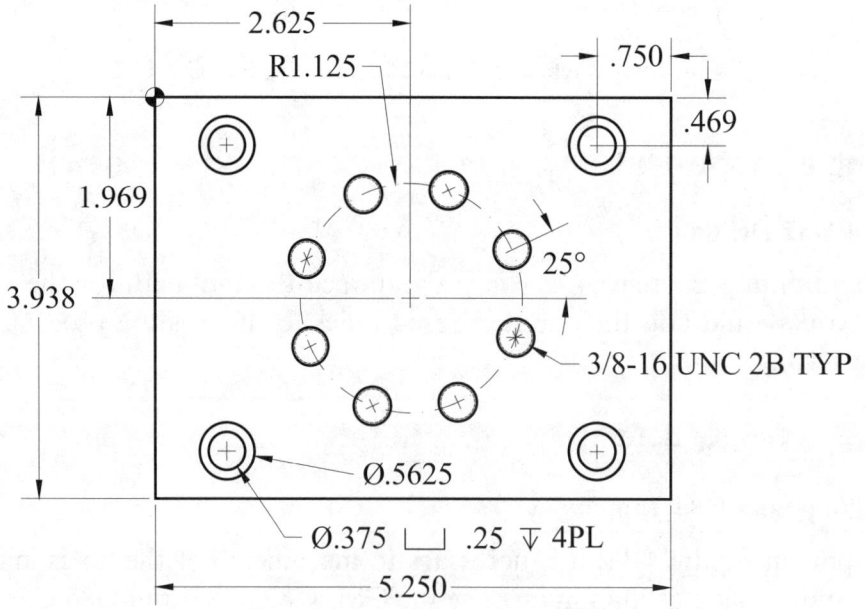

Figure 4-11   Machining Center Programming Exercise 4-14

CNC Machining Center Combined Projects

Programming Exercise 4-15

In the Process Planning section of this workbook, you identified the operation, tools, and setup information for the part in Figure 4-12. Now use this information that you gathered to write a program for the same part. Note: the step may be axially roughed to within .030 inch with the face mill; also, the top surface must have .125 inch material removed. In this case, the tool path should be programmed without the use of Cutter Radius Compensation (program the exact tool path) for the face mill.

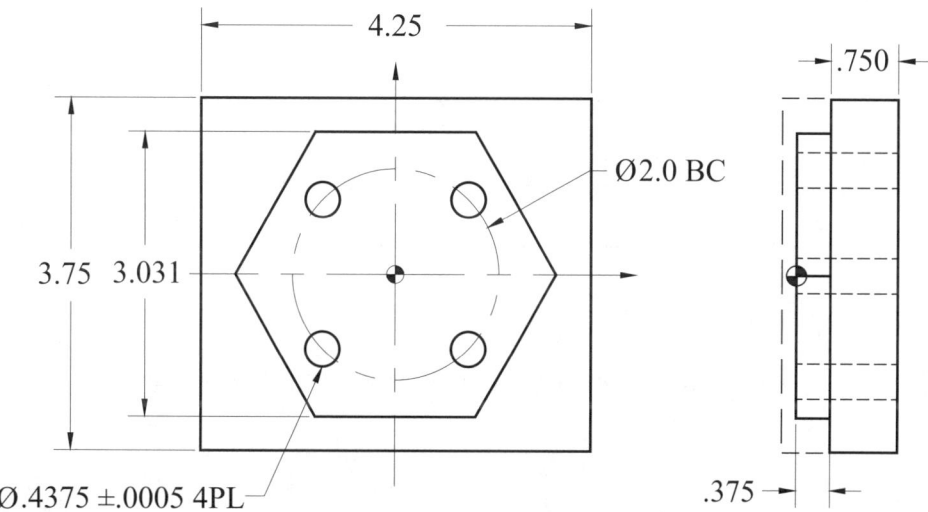

Figure 4-12   Machining Center Programming Exercise 4-15

CNC Machining Center Subprogram Application

Programming Exercise 4-16

M98, M99, and G68

In this project, several semicircular slots of the same size need to be machined in the part (Figure 4-13). These are excellent candidates for subprogram application. Program one of the slots as a subprogram and call it multiple times from within the main program. Also, the use of multiple work offsets will accommodate this. In this case, the part is aluminum and is 3/4-inch thick.

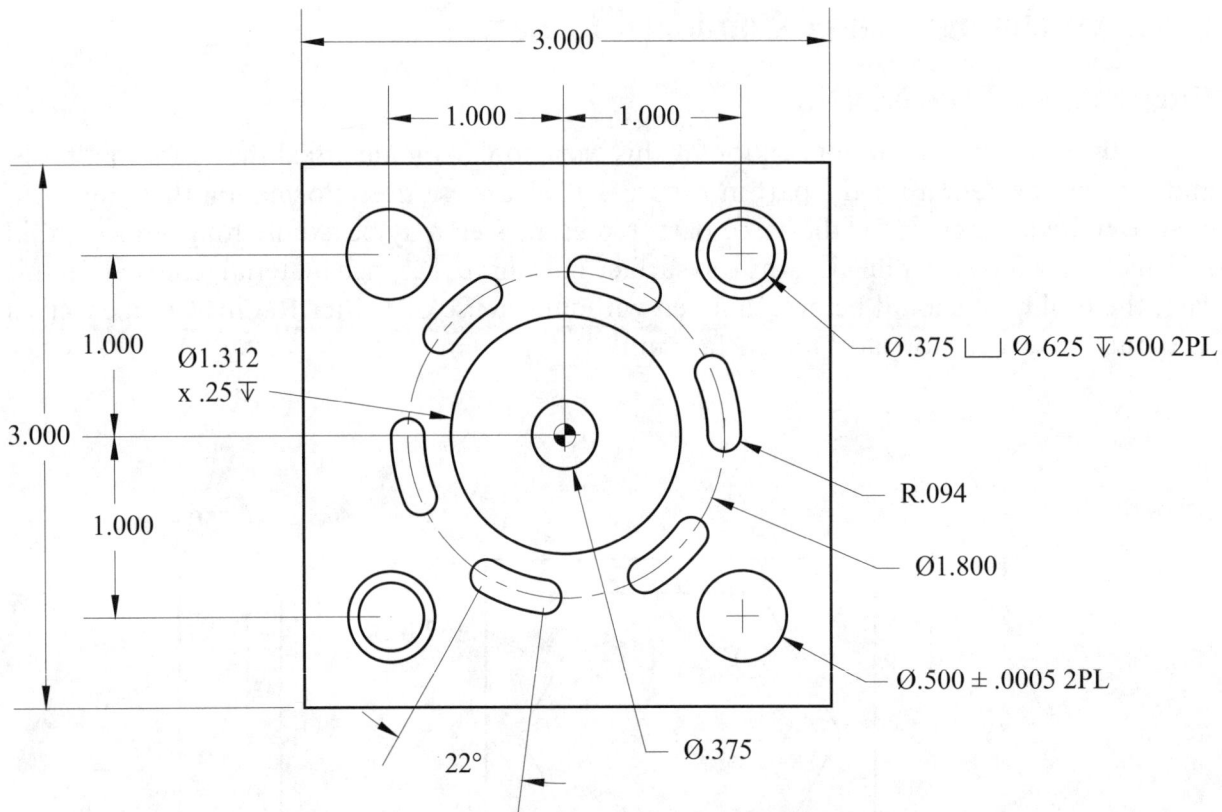

Figure 4-13   Machining Center Programming Exercise 4-16

Programming Exercise 4-17

G98

In this example, it is necessary to machine all of the holes and step cutouts in the block in Figure 4-14. The material is carbon steel. Where possible, use the G98 command with Canned Cycles to make the program. The part is already machined to the overall size and thickness dimensions.

CNC Machining Center Program Error Diagnosis

Use the skills you have learned to identify the problems in the following program lines and program sections. You may refer to the text, *Programming of CNC Machines*, Fourth Edition.

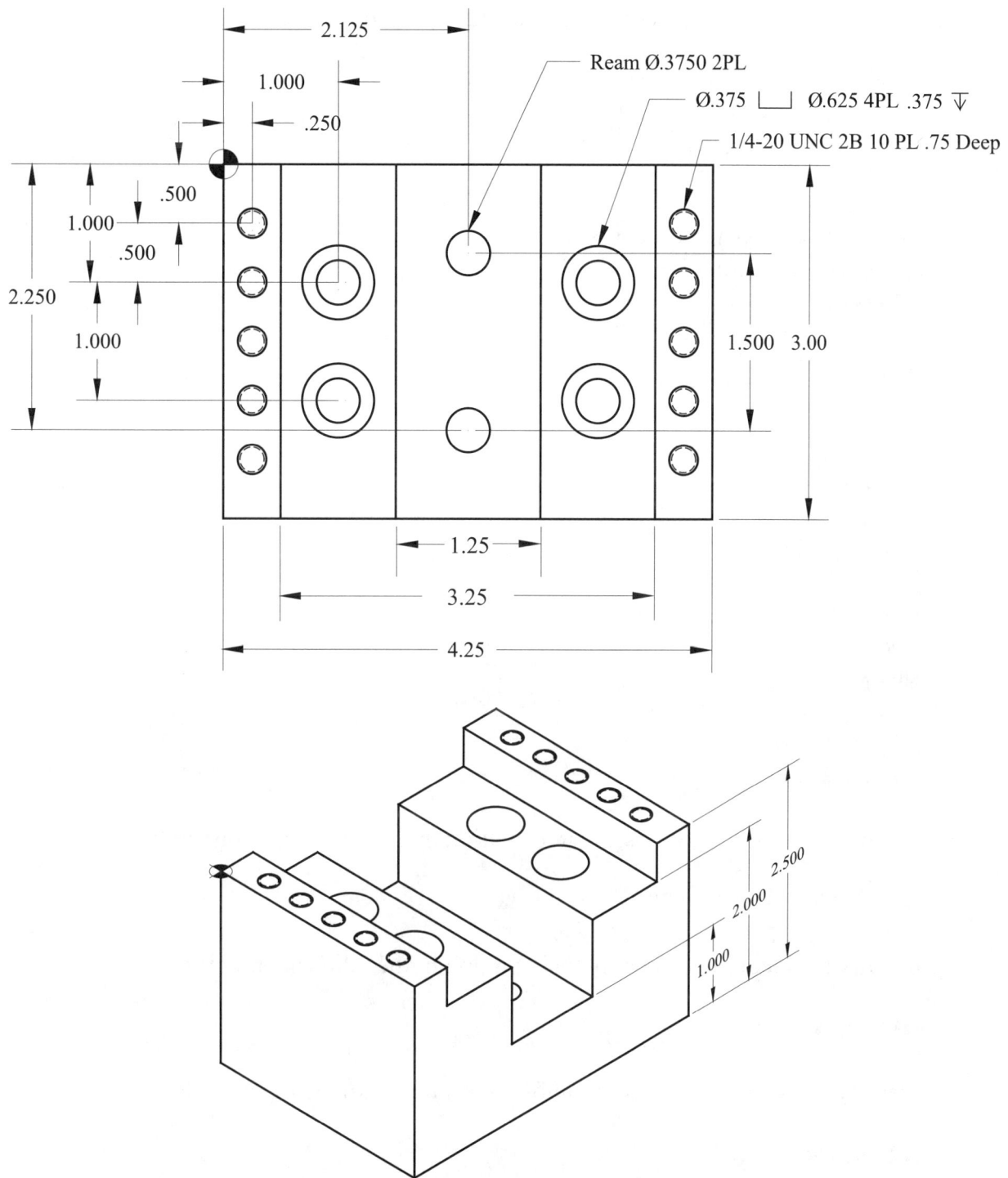

Ream Ø.3750 2PL

Ø.375 ⌴ Ø.625 4PL .375 ▽

1/4-20 UNC 2B 10 PL .75 Deep

Figure 4-14, Machining Center Programming Exercise 4-17

1. Use the CNC code and sketch a representation of the part being created.

   O3001
   N10 G90 G17 G20 G80 G49
   N15 G28 X0.0 Y0.0 Z0.0
   (3/8 2 FLUTE ENDMILL)
   N20 T1 M6
   N25 G00 G54 X0.0 Y0.0 S1426 M3
   N30 G43 H1 Z.1 M8
   N35 G1 Z–.1 F6.33
   N40 G41 D1 Y6.
   N45 X1.0 Y7.0
   N50 X1.5
   N55 X2.5 Y6.0
   N60 G3 X4.5 R1.0
   N65 G1 X11.0 Y2.5
   N70 Y1.0
   N75 X10.0 Y0.0
   N80 X0.0
   N85 G40 X–.2
   N90 Z.1
   N95 M5
   N100 G91 G0 G28 Z0.0 M9
   N105 G28 X0.0 Y0.0
   N110 M30

2. Identify the incorrect or missing information in the following program line.

   N250 G02 X–.375 Y2.938

3. Identify the incorrect or missing information in the following program line.

   N110 X–4.625 F12.0

4. Identify the incorrect or missing information in the following program line.

   N25 S2500 M4

5. Identify the missing information in the following program line.

   N125 G83 G99 Z–1.13 R.1 F3.6

6. Identify the missing or incorrect information in the following program and/or subprogram.

    O2010
    N10 G90 G80 G20 G40 G49 M23
    N15 G00 G54 X–1.25 Y.75 S1000 M03
    N20 G43 Z1.0 H01 M08
    N25 G81 G98 Z–.35 R.1 F6.0
    N30 M98 P7
    N35 G00 G80 X0.0
    N40 M21
    N45 G00 X–1.25 Y.75
    N50 G81 G98 Z–.35 F6.0 R.1
    N55 M98 P7
    N60 G00 G80 X0.0 Y0.0
    N65 M23
    N70 M22
    N75 G00 X–1.25 Y.75
    N80 G81 G98 Z–.35 F6.0 R.1
    N85 M98 P7
    N90 G80 G00 X0.0 Y0.0
    N95 M21
    N100 G00 X–1.25 Y.75
    N105 G81 G98 Z–.35 F6.0 R.1
    N110 M98 P7
    N115 G80 Z1.0 M09
    N120 M23
    N125 G91 G28 X0.0 Y0.0 Z0.0
    N130 M30

    Subprogram for Program 2011
    O2011
    N1 X–2.5
    N2 X–3.75
    N3 Y1.5
    N4 X–2.5
    N5 X–1.25
    N6 Y2.25
    N7 X–2.5
    N8 X–3.75
    N9 M30

7. Identify the missing information in the following program line.

   N310 G01 G41 X–4.125 Y0.0

8. Identify the missing information in the following program line.

   N10 G90 G20 G80 G49

9. Identify the missing information in the following program line.

   N35 G43 Z1.0

Answer Key
Answers to Unit 1: CNC Basics

Process Planning Answers

Turning Center Project

Review the Operation Sheet (Chart A1-1), the CNC Setup Sheet (Chart A1-2), and the Quality Control Check Sheet (Chart A1-3) for the turning center project and compare your answers. Figure 1-1 is repeated here to help with your review.

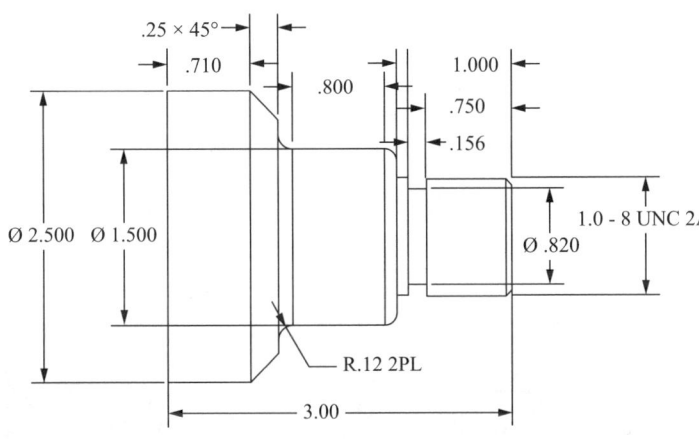

Figure 1-1   Turning Center Process Planning

Chart A1-1 Turning Center Operation Sheet

Date: Today	Prepared By: You	
Part Name: Turning Center Project	Part Number: 1234	
Quantity: 1	Sheet _1_ of _1_	
Material: Alloy Steel		
Raw Stock Size: 12' x 2.50" diameter bar stock		

Operation Number	Machine Used	Description of Operation	Time
1	Saw	Cut the bar stock to 3.0625" lengths	
2	CNC Turning Center	Machine complete to dimensions	
3	Bench	Deburr as needed	
4	QC	Final 100% Quality Control inspection	

Turning Center
Chart A1-2 CNC Setup Sheet

Date: Today	**Prepared By:** You
Part Name: Turning Center Project	**Part Number:** 1234
Machine: CNC Turning Center	**Program Number:** TBD

Workpiece Zero: **X** _Centerline_   **Y** _NA_   **Z** _Finished Face_

Setup Description:

Clamp the part in a 3-jaw chuck with soft-jaws, 2.35" minimum extended out of the chuck.

Tool Number	Offset Number	Tool Description	Comments
02	02	Rough Turning Tool .031 TNR	SFPM 125–1000
04	04	Finish Turning Tool .015 TNR	SFPM 125–1000
06	06	OD Grooving Tool	.118 Wide .005 TNR
08	08	OD Threading Tool	SFPM 125–1000

Facing Cut = r/min 191–1527
Turning 1.0" diameter, r/min = 477–3820
Turning 1.5" diameter, r/min = 318–2546
Turning 1.0" diameter, r/min = 191–1528
Turning 1.0" diameter, r/min = 582–4658
Threading r/min = 477–3820 Feed = Thread Lead
NOTE: The preferred method of setting the r/min for threading is to use the Constant Cutting Speed (G96) command.

Turning Center
Chart A1-3 Quality Control Checksheet

Date: Today			Checked By: You

Part Name: Turning Center Project			Part Number: 1234

			Sheet _1_ of _1_

Blueprint Dimension	Tolerance	Actual Dimension	Comments
Ø 2.50	±.010		
Ø 1.50	±.010		
.75	±.010		
.25	±.010		
45°	±.5°		
R .10 2PL	±.015		
1.0 - 8 Thread	.9980/.9830		
1.0" - 8 Thread Pitch Diameter	.9168/.9100		
.05	±.010		
45°	±.5°		
.75	±.010		
.156	±.005		
.09	±.010		
3.00	±.010		

Machining Center Project

Review the Operation Sheet (Chart A1-4), the CNC Setup Sheet (Chart A1-5), and the Quality Control Check Sheet (Chart A1-6) for the machining center project and compare your answers. Figure 1-2 is repeated here to help with your review.

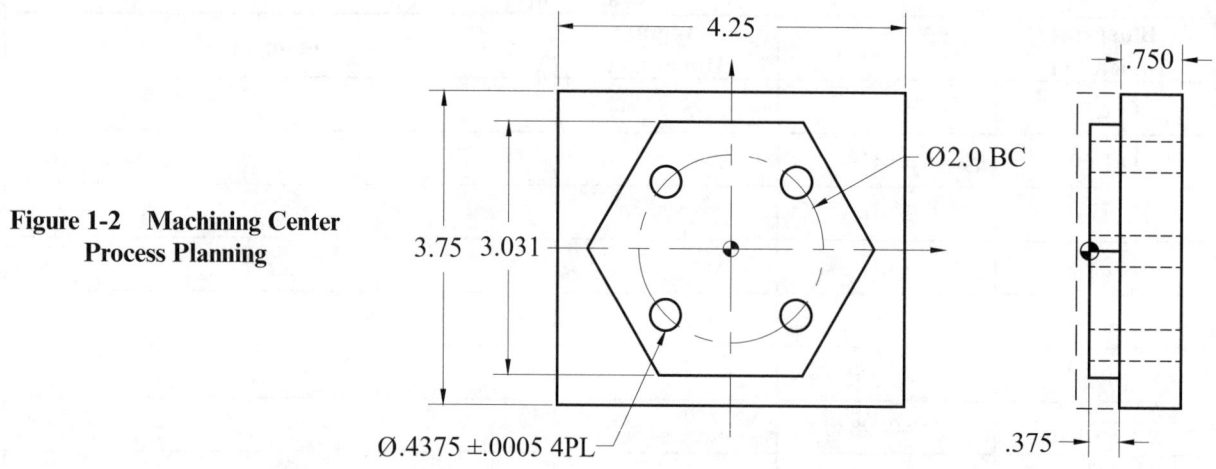

**Figure 1-2   Machining Center
Process Planning**

Chart A1-4 Machining Center Operation Sheet

Date: Today	**Prepared By:** You
Part Name: Machining Center Project	**Part Number:** 1235
Quantity: 1	**Sheet 1 of 1**
Material: Aluminum	
Raw Stock Size: 4.5" x 1.25" bar stock	

Operation Number	Machine Used	Description of Operation	Time
1	Saw	Cut the bar stock to 3.875" lengths	
2	Mill	Machine blanks to 4.25" x 3.75" finished size	
3	CNC Machining Center	Machine complete to dimenstions	
4	Bench	Deburr as needed	
5	QC	Final 100% Quality Control inspection	

Machining Center
Chart A1-5 CNC Setup Sheet

Date: Today	Prepared By: You
Part Name: Machining Center Project	Part Number: 1235
Machine: CNC Machining Center	Program Number: TBD

Workpiece Zero: X _Centerline_ Y _Centerline_ Z _Top Most Finished Surface_

Setup Description:

Clamp the part in a vise on parallels.
A minimum of 1/2" must be above the vise jaws.

Tool Number	Offset Number	Tool Description	Comments
1	1	Ø 3.0" 5 tooth face mill	SFPM 755–1720 in/tooth = .020–.039
2	2	Ø .75" 2-Flute End Mill	SFPM 165–850 in/tooth = .002–.006
3	3	Ø .50" x 90° Spotting Drill	SFPM 165–850 in/tooth = .002–.006
4	4	Ø .4219" 27/64 Drill	SFPM 165–850 in/tooth = .002–.006
5	5	Ø 4375" Reamer	r/min = 1/3 of lowest drill speed feed = 1/2 of lowest drill feed

Tool 1 = r/min 961–2190 = ipm 96–427
Tool 2 = r/min 840–4329 = ipm 17.32–51.9
Tool 3 = r/min 1260–6494 = ipm 2.52–38.96
Tool 4 = r/min 1494–7696 = ipm 5.98–92.35

Answers to Unit 1: CNC Student Workbook

Machining Center
Chart A1-6 Quality Control Checksheet

Date: Today		Checked By: You	
Part Name: Machining Center Project		Part Number: 1235	
		Sheet 1 of 1	
Blueprint Dimension	Tolerance	Actual Dimension	Comments
4.25	±.010		
Ø 3.0 BC	±.015		
.4375	±.0005		
.4375	±.0005		
.4375	±.0005		
.4375	±.0005		
3.75	±.010		
3.031	±.010		
3.031	±.010		
3.031	±.010		
.375	±.005		
.750	±.005		

Feeds and Speeds Answers

1. Alloy Steel SFPM = 125–1000
   r/min (RPM) = at the largest diameter (2.5) and the slowest CS,
   r/min = 191 and 1527 at the largest diameter and highest CS
   in/rev =.008–.036
   D = 2.5 to 0

Because in facing the diameter reaches zero eventually, the r/min must change as the diameter changes. On a CNC lathe, the use of Constant Surface Speed is recommended (G96 code). To accomplish a uniform SFPM, the r/min attained is limited by the max RPM of the Machine Tool.

Because the amount of material to be removed is only 1/32, it would also be the acceptable amount to use for a depth of cut. Depending on the surface finish requirement, it may also be acceptable to take 2 cuts at .0156 each.

See *Programming of CNC Machines*, Fourth Edition, Part 3 (Programming CNC Turning Centers, Constant Cutting Speed).

2. Aluminum SFPM = 2800–4500
   r/min = 4625–7433
   Feed or in/rev = .017–.036

3. Carbon Steel SFPM = 30–160 for HSS tool
   r/min = 115–611
   Feed or in/rev = .006–.012

4. See the Turning Center CNC Setup sheet (Chart A1-2) and compare your answers.

5. See the Machining Center CNC Setup sheet (Chart A1-5) and compare your answers.

6. Stainless Steel SFPM = 20–80
   r/min = 142–569
   in/tooth .001–.003
   in/min = .284–3.4

7. Carbon Steel SFPM = 25–140
   Drill r/min = 80–448 and in/min = .160–3.58
   End Mill r/min = 114–640 and in/min = .456–10.24

8. Alloy Steel SFPM = 39–475
   r/min = 52–633
   in/tooth = .020–.039
   in/min = 5.2–123.43

9. Aluminum SFPM = 165–850 for an HSS cutter
   r/min = 1508–7770
   in/tooth = .002–.006
   in/min = 6.03–93.24

10. Check your answers with the speeds and feeds listed before each of the programs.

Coordinate Systems Answers

1. Absolute coordinates for each axis and for each point of the profile of the turned part, based on diametrical considerations. Refer to Figure 1-3.
   X.90 Z0.0, X1.0 Z–.05, X1.0 Z–.75, X.91 Z–.75, X.82 Z–.906, X1.0 Z–.906, X1.0 Z–1.0, X1.3 Z–1.0, X1.5 Z–1.1, X1.5 Z–1.9, X1.7 Z–2.0, X2.0 Z–2.0, X2.5 Z–2.25, X2.5 Z–3.0

2. Incremental coordinates for each axis and for each point of the profile of the turned part, based on radial considerations. Refer to Figure 1-3.
   X0.0 Z0.0, X.450 Z0.0, X.05 Z–.05, X0.0 Z–.70, X–.045 Z0.0, X0.0 Z–.156, X.09 Z0.0, X0.0 Z–.094, X.15 Z0.0, X.1 Z–.1, X0.0 Z–.8, X.1 Z–.1, X.25 Z0.0, X.25 Z–.25, X0.0 Z–.75

3. Absolute coordinates for each axis and for each point of the profile of the milled part. Start at part zero and proceed clockwise. Refer to Figure 1-4.
   X0.0 Y0.0, X0.0 Y6.0, X1.0 Y7.0, X1.5 Y7.0, X2.5 Y6.0, X4.5 Y6.0, X11.0 Y2.5, X11.0 Y1.0, X10.0 Y0.0

4. Incremental coordinates for each axis and for each point of the profile of the milled part. Refer to Figure 1-4.
   X0.0 Y0.0, X0.0 Y6.0, X1.0 Y1.0, X.5 Y0.0, X1.0 Y–1.0, X2.0 Y0.0, X5.5 Y–3.5, X0.0 Y–1.5, X–1.0 Y–1.0, X–10.0 Y0.0

5. Absolute coordinate values for X, Y, and Z for each of the 15 points as indicated on the following drawing. Refer to Figure 1-5.

1. X0.0 Y0.0 Z0.0	**6. X0.0 Y–2.0 Z0.0**	**11. X–3.0 Y–.75 Z0.0**
2. X0.0 Y–.75 Z0.0	**7. X–3.0 Y–2.0 Z0.0**	**12. X–3.0 Y0.0 Z0.0**
3. X0.0 Y–.75 Z–.375	**8. X–3.0 Y–1.25 Z0.0**	**13. X–3.0 Y0.0 Z–2.25**
4. X0.0 Y–1.25 Z–.375	**9. X–3.0 Y–1.25 Z–.375**	**14. X–3.0 Y–2.0 Z–2.25**
5. X0.0 Y–1.25 Z0.0,	**10. X–3.0 Y–.75 Z–.375**	**15. X0.0 Y–2.0 Z–2.25**

6. Identify each axis (vertical milling representation) and its positive or negative value in Figure 1-6. (See Figure A1-1.)

7. Indicate the negative rotation direction for the polar coordinate system in Figure 1-6. (See Figure A1-1.)

8. Indicate each of the polar quadrants in Figure 1-6. (See Figure A1-1.)

9. Identify the angular value locations for 0, 90, 180, and 270 degrees in Figure 1-6. (See Figure A1-1.)

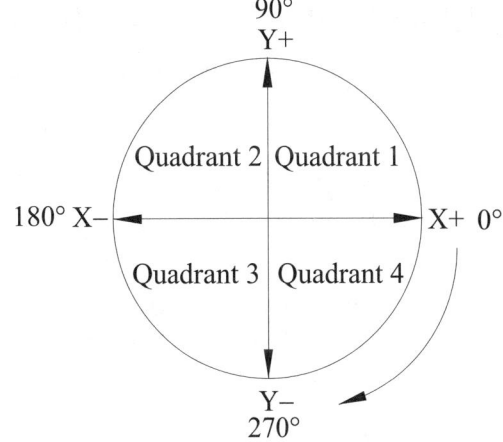

Figure A1-1 Answer for Vertical Milling Axes, Polar Rotation, Quadrants, and Angular Values

10. Identify the polar (angular and radial) values for each of the holes in Figure 1-7. (See Figure A1-2.)

11. Identify the polar (angular and radial) values for each of the holes in Figure 1-8. (See Figure A1-3.)

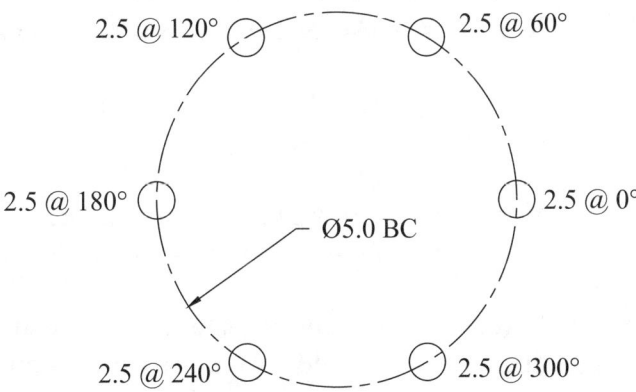

Figure A1-2 Answer for Polar Coordinate Values

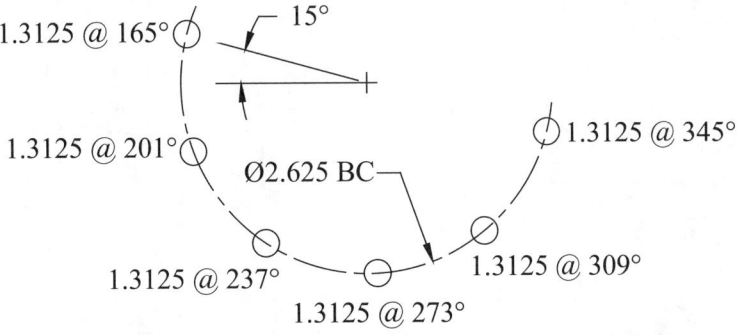

Figure A1-3 Answer for Polar Coordinate Values

Trigonometric Calculations Answers

1. Absolute rectangular coordinate locations for each hole center point. (See Figure A1-4.)

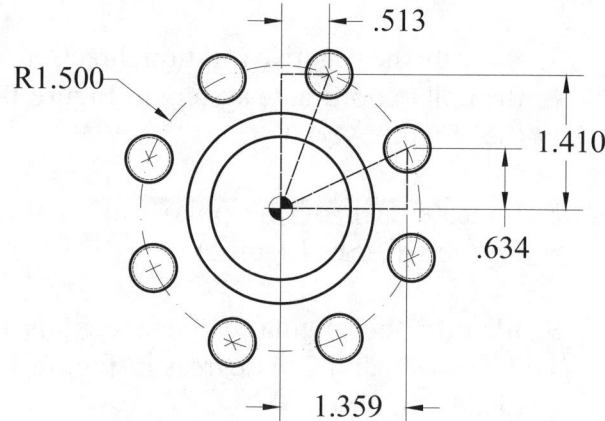

**Figure A1-4   Answers for Calculate
Absolute Coordinates**

**X.513Y1.410, X1.359Y.634, X1.359Y–.634, X.513Y–1.410, X–.513Y–1.410,
X–1.359Y –.634, X–1.359Y.634, X–.513Y1.410**

2. Center-to-center dimension. Use the Pythagorean Theorem formula to calculate. (See Figure A1-5.)

3. Value for the unknown chord distance. Use the Pythagorean Theorem formula to calculate. (See Figure A1-6.)

4. Values for each chord distance. Using the Oblique triangle formulas and the values for a known side length and a known angle, calculate the chord values. (See Figure A1-7.)

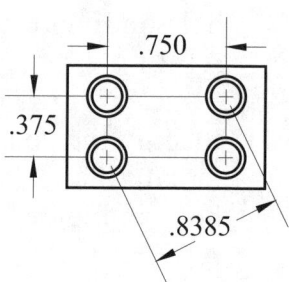

**Figure A1-5   Answer for Calculate
Center-to-Center Distance**

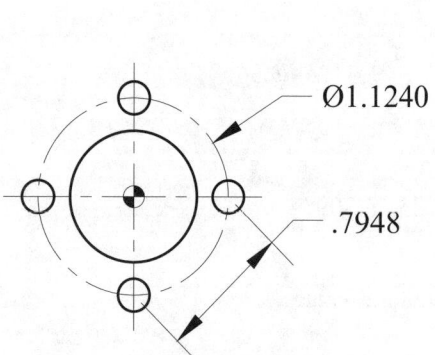

**Figure A1-6   Answer for Calculate
Center-to-Center Distance**

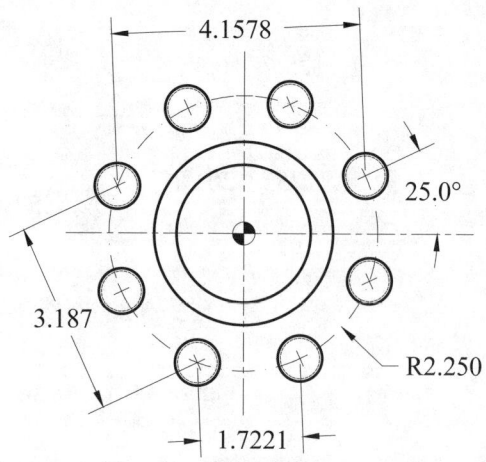

**Figure A1-7   Answers for Calculate
Chord Distances**

5. Tool travel necessary to allow for the drill point. (See Figure A1-8.)
   .437/2 = .2185
   .2185 × TAN 31° = .1313
   .562 + .1313 = .6933

6. Depth of cut required to countersink to a diameter of .395 inch. (See Figure A1-9.)
   .0938 × TAN 30° = .054 (for the 120° tip)
   The length from the 120° tip to the start of the 60° portion = .1875
   .395 diameter − .1875 diameter/2 = .1038
   .1038 × TAN 60° = .1798 (for the 60° portion)
   .054 + .1798 + .1875 = .4213

7. Tool travel necessary to allow for the drill point plus .090. (See Figure A1-10.)
   .090 + .875 + .0938 (drill point) = 1.0588

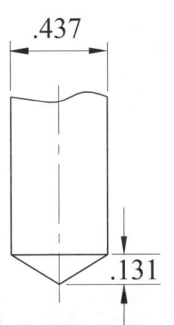

**Figure A1-8 Answer
for Calculate Drill Point
Compensation**

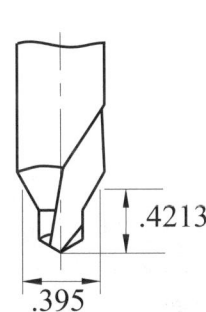

**Figure A1-9 Answer
for Calculate Drill Point
Compensation**

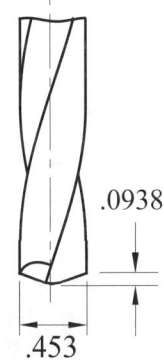

**Figure A1-10 Answer
for Calculate Drill Point
Compensation**

8. Offset amount for each axis and the coordinate values that will be required for the CNC program. (See Figure A1-11.)

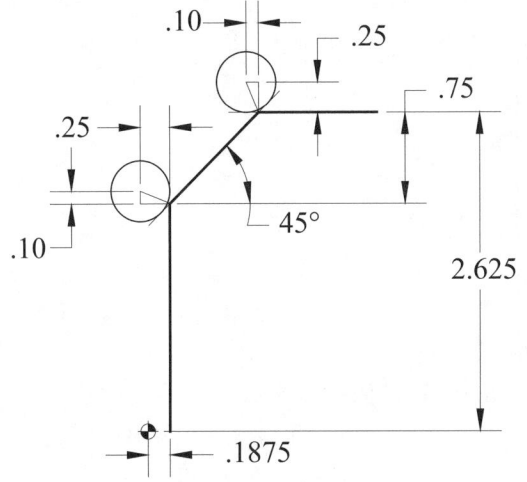

**Figure A1-11 Answer for Calculate Cutter
Offset Coordinates**

9. Offset for the tool nose radius when turning a 30° tapered surface. (See Figure A-12.)

10. In Figure 1-18, a calculation is necessary to offset for the tool nose radius when turning a 30° tapered surface. The face and centerline of the turned part are zero. List the coordinates needed in the CNC program to allow for this offset. (See Figure A1-13.)

11. Offset amount for the tool nose radius when turning a .062 inch 45° chamfer. (See Figure A1-14.)

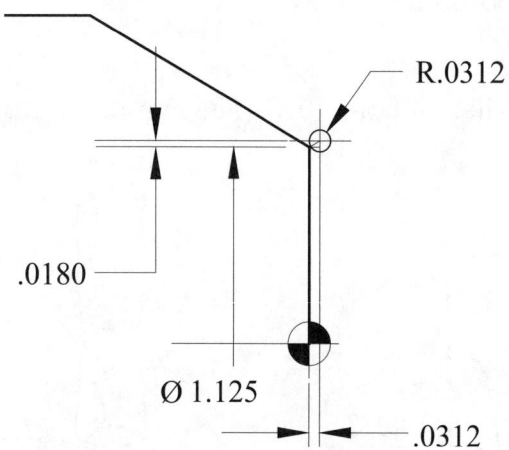

Figure A1-12   Answer for Calculate Tool Nose Radius Offset Coordinates

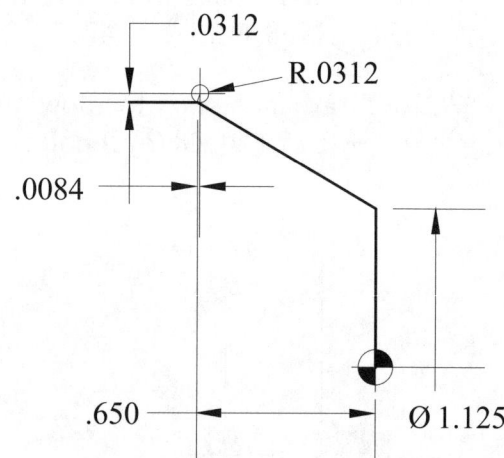

Figure A1-13   Answer for Calculate Tool Nose Radius Offset Coordinates

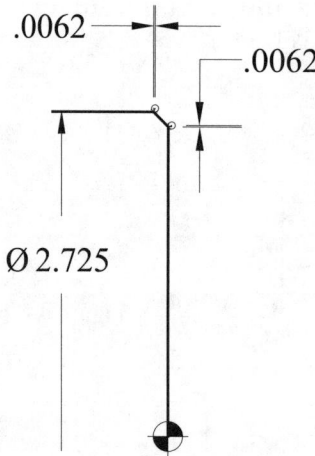

Figure A1-14   Answer for Calculate Tool Nose Radius

CNC Basics Study Question Answers

Answers are indicated in **BOLD** letters.

1. Programming is a method of defining tool movements through the application of numbers and corresponding coded letter symbols.

    T or F

2. A lathe has the following axes:
    a. X, Y, and Z
    b. X and Y only
    c. X and Z only
    d. Y and Z only

3. Program coordinates that are based on a fixed origin are called:
    a. Incremental
    b. **Absolute**
    c. Relative
    d. Polar

4. On a two-axis turning center the diameter controlling axis is:
    a. B
    b. A
    c. X
    d. Z

5. The letter addresses used to identify axes of rotation are:
    a. U, V, and W
    b. X, Y, and Z
    c. A, Z, and X
    d. **A, B, and** C

6. The acronym TLO stands for:
    a. **Tool Length Offsets**
    b. Total Length Offset
    c. Taper Length Offset
    d. Time Length Offset

7. When referring to the polar coordinate system, the clockwise rotation direction has a positive value.

    T or **F**

8. In the text *Programming of CNC Machines*, Figure 1-15 of Part 1, in which quadrant is the part placed in?
    3

9. A program block is a single line of code followed by an end-of-block character.
    T or F

10. Each block contains one or more program words.
    T or F

11. Using Figure 1-15, Part 1, *Programming of CNC Machines*, Fourth Edition, list the X and Y absolute coordinates for the part profile where workpiece zero is at the lower left corner. (The corner cutoff is at a 45° angle).
    X0.0Y0.0, X0.0Y5.0, X2.5Y5.0, X4.0Y3.5, X4.0Y0.0, X0.0Y0.0

12. Using Figure 1-15, Part 1, *Programming of CNC Machines*, Fourth Edition, list the X and Y incremental coordinates for the part profile where workpiece zero is at the lower left corner.
    X0.0Y0.0, X5.0Y2.5, X1.5Y–1.5, Y–3.5, X–4.0

13. How often should machine lubrication levels be checked?
    Daily

Answer Key
Answers to Unit 2: Setup and Operation

General Steps Answers

1. General steps required in preparing a CNC machine for production of a programmed part that has been effectively run before:
   - Study the blueprint.
   - Check the raw material requirements and verify compliance.
   - Study any Planning Documents (e.g., Operation, CNC Setup, and Quality Control sheets) supplied.
   - Collect all required work holding and cutting tooling.
   - Identify the program # and load it into the active controller memory.
   - Install all work holding and cutting tools to their correct locations.
   - Align (dial-in) work holding.
   - Measure Tool Length Offsets (TLO).
   - Measure Work Offset.
   - Perform a tool path verification of the program on the control, where available.
   - Perform a dry run of the program cycle either without a part mounted or above the finished surface by some amount (commonly 1.0 inch) to physically verify machine movements and tool usage.
   - Remove any Z-axis offset used for the dry run procedure.
   - Set the rapid traverse to a reduced level.
   - Set the Optional Stop to an ON condition.
   - Use Single-Block operation for the first several moves to verify proper tool approach.
   - Automatic operation.
   - Perform in-process inspection at Program or Optional Stops and adjust where necessary.
   - Complete the program cycle.
   - Perform a 100% dimensional check of the part.
   - Production may begin after adjustments are made verified and a consecutive inspection is acceptable.

Operation Scenario Answers

2. Pulse Generator (Handle) magnitude increments.
    X1 = a movement of .0001 inch or .0025 millimeters (mm)
    X10 = a movement of .001 inch or .0254 mm
    X100 = a movement of .010 inch or .254 mm
    X1K = a movement of .100 inch or 2.54 mm
(See *Programming of CNC Machines*, Fourth Edition, Part 2: CNC Machine Operation.)

3. What will occur when the reset button is pressed during automatic operation? What steps should be followed to recover?

   • When the reset button is pressed during the automatic cycle, all feed movement will halt, spindle rotations will halt, and coolant flow will halt. The program will be reset to the program beginning. In order to recover from this condition, the operator must switch to a manual mode (e.g., Jog or Handle) and set the appropriate axis switch (commonly Z-axis) for removal of the tool that is in-cut, if necessary.

   • Move the tool to a clear position.

   • Home the machine or move it to a location where enough room is allowed to repair any problem that exists.

   • Replace or repair any broken tooling and re-measure the TLO.

   • Set the mode to EDIT and use the cursor or other search method to return the program to the beginning of its sequence for the tool in use, or press the reset button to set the program at its beginning, if applicable.

   • Return to the Automatic Operation mode and press cycle start. The rapid traverse override may be set to a lower level and the Single Block mode activated to ensure that proper positioning is attained.

4. What differences would exist if the Emergency Stop button is pressed during automatic operation rather than the reset button? What will occur? What steps should be followed to recover?

   • All of the same results will occur. The difference will be in the recovery. First a release of the E-Stop button will be required. Do so by twisting it in the clockwise direction until it pops back out. On some older machines (check the Operation Manuals supplied with the machine), the machine must be "Homed" prior to returning to automatic operation. The other steps are identical as with the reset button condition.

5. When the Feed Hold button is pressed during automatic operation, what will occur? What steps should be followed to recover?

   • When the Feed Hold button is pressed, only feed movement is stopped. The Spindle remains ON and coolant flow is not interrupted. Merely press the Cycle Start button again to resume automatic operation.

6. What mode of operation is required to install a tool into the spindle by the Automatic Tool Changer (ATC)?

> a. Automatic
> b. Jog
> c. Edit
> **d. Manual Data Input (MDI)**

7. What is the appropriate and safe method for performing in-process inspection measurements?

- If a program stop (M00) is in the program before each tool change, measurements can be taken at this time.

- If an optional stop (M01) is in the program before each tool change and the Optional Stop button is activated, measurements can be taken at this time.

- Another less popular method is to wait until the program is completed, and then measurements can be taken before the part is removed from the work holding device.

8. Where and how is the change made to compensate for a .003 inch diametrical variation for tool #4?

- To make the adjustment on the offset page, position the cursor under the wear column for tool offset #4 and input an amount of minus .003 inch. Press the INPUT key.

9. Clearing of an existing wear offset for tool #4.

- Position the cursor to tool #4's wear offset and use the numeric keypad to enter a value of zero; then press the INPUT button. Or, position the cursor to tool #4's wear offset and use the numeric keypad to enter a value that is opposite the value shown (e.g., if the value is .003, then key in –.003); then press the INPUT+ button.

10. Restarting the program from a specific tool # from that point on.

In the EDIT mode, use the word search method to search to a specific word in the program, (e.g., T6) and follow these directions:

- From the EDIT mode, use the alphanumeric keypad key in the letter address T.
- Press the number 6.
- Press the SRH soft key forward or reverse for the direction needed.
- The cursor will move to the identified word T6.
- Re-enter the automatic execution mode and press Cycle Start.

MDI Usage Answers

11. Several tasks that can be accomplished by using Manual Data Input MDI.

   - Turning the spindle ON, in a clockwise direction, at r/min. For the mill the code is S1000M3, and for the lathe the code is G97S500M3.

   - Executing a tool change, for the mill, T1M6, and for the lathe, T0101.

   - Turning ON the flood coolant flow, M08.

   - Turning OFF the coolant flow, M09.

   Other common tasks that can be performed are:

   - The use of absolute rapid positioning moves that are based on a specific work offset, for example, G90 G00 G54 X0.0 Y0.0

   - The execution of a tool offset to verify proper tool length measurement accuracy.

   - Using G43 Z1.0 H01 for milling or T0101 Z.1 for turning.

   - Homing the machine by using the reference return command for milling G91 G28 Z0.0 and G28 X0.0 Y0.0, or for turning G28 U0.0 W0.0.

   - The execution of incremental feed moves for cutting vise jaws, etc., G91 G01 X–10.0 F20.0.

   - Any command that can be a part of the program can be input and executed by using MDI. However, there is a limit of 10 lines of consecutive input in this mode. (See *Programming of CNC Machines*, Fourth Edition, Part 2: CNC Machine Operation.)

Programming Editing Answers

12. Altering of a program word G55 to G54 in an existing program.

   - From the EDIT mode, use a searching method to scan the program to the word to be altered.

   - Use the alphanumeric keypad to key the new address and the new data to be inserted.

   - Press the ALTER key.

   - The new data are changed.

   To change the program word G55 in the example to G54, follow these steps:

   - Press the EDIT key.

   - Press the PRGRM soft key.

   - Key in the program word, G55.

   - Press the SRH soft key in the forward direction.

- Key in the new program word to insert G54.
- Press ALTER.

This procedure must be repeated to ensure that all instances of the incorrect offset calls are corrected. When using an offline editing software, a common method for doing this is called "find and replace".

13. Insertion of a missing program word into an existing program.
- From the EDIT mode, use a searching method to scan the program to the word immediately before the word to be inserted.
- Use the alphanumeric keypad to key the address and the data to be inserted.
- Press the INSERT key.
- The new data is inserted.

Example: To insert the program word T0101 on sequence number N15 of the program listed below:

- Press the EDIT key
- Press the PRGRM soft key
- Key in the program word, S1000
- Press the SRH soft key in the forward direction
- Key in the new word to insert N2 T0101
- Press INSERT.
    O1234
    N10 G50 S1000
    N20 G96 S600 M03
    N25 G00 X1.2 Z.2

14. Deleting a program word from an existing program.
- From the EDIT mode, use a searching method to scan the program to the word that needs to be deleted.
- Then press the DELETE key.

To delete the program word (Date 03/29/15) from the example, follow these steps:
- Press the EDIT key
- Press the PRGRM soft key
- Key in the word (Date 03/29/15)
- Press the SRH soft key in the forward direction
- Press DELETE.

Setup and Operation Study Question Answers

Answers are indicated in **bold** letters.

1. The counterclockwise direction of rotation is always a negative axis movement when referring to the handle/pulse generator.
    T or F

2. Which display includes the programmed Distance-to-Go readouts?
    Program Check

3. When the machine is ON and the program check screen is displayed, there is a list group of G-Codes displayed. What does this indicate?
    Default codes that are active

4. Describe the difference between the Input and the +Input soft keys in the function.
    The Input key, when used, inputs a whole number.
    The Input + key, when used, inputs an incremental amount.

5. Which button is used to activate automatic operation of a CNC program?
    a. Emergency Stop
    b. Cycle Stop
    c. Cycle Start
    d. Auto

6. Which display lists the CNC program?
    a. Position page
    b. Offset page
    c. Program check
    d. Program page

7. When the machine is turned on for the first time, it must be sent to its home position.
    T or F

8. Which operation selection button allows for the execution of a single CNC command?
    a. Dry run
    b. Single block
    c. Block delete
    d. Optional stop

9. Which mode switch/button enables the operator to make changes to the program?
    a. Edit
    b. MDI
    c. Auto
    d. Jog

10. What does the acronym MDI stand for?
    Manual Data Input

11. Which display screen is used to enter tool information?
    Offset

12. If the Reset button is pressed during automatic operation, then spindle rotations, feed, and coolant will stop.
    T or F

13. During setup, the mode switch used to allow for manual movement of the machine axes is:
    a. Auto
    b. MDI
    c. Edit
    d. Jog

Answer Key

Answers to Unit 3: CNC Turning Center Programing

The answer programs given here are only one method of machining the parts shown in the figures. There are many possibilities for getting the correct results. Where possible, tool path verification and simulation programs should be used to confirm program integrity. Students should also consult with their instructor for program verification and always proceed, with caution, when running a newly written program. In the following exercises where only one tool is required, a CNC Setup sheet is not necessary; however, the tool and setup information should be listed before your program code. Please use a CNC Setup sheet in all other cases for the sake of clarity.

Identifying Programming Coordinates for Turning Answers

1. Absolute programming coordinates including arc center locations with X-axis radial values. Refer to Figure 3-1.
   X0Z0, X.3Z0, (Arc Center = X.3Z–.2) X.5Z–.2, X.5Z–.383,
   (Arc Center = X1.261Z–.697) X1.2Z–1.521, X1.2Z–1.978,
   (Arc Center = X2.197Z–2.770) X1.25Z–3.745,
   (Arc Center = X1.625Z–3.775) X1.625Z–4.150, X1.80Z–4.150,
   (Arc Center = X1.80Z–4.35) X2.0Z–4.35, X2.0Z–4.7

2. Incremental programming coordinates including arc center locations with X-axis radial values. Refer to Figure 3-1.
   X0Z0, U.3W0, (Arc Center = U0W–.2) U.2W–.2, U0W–.183,
   (Arc Center = U.761W–.314) U.7W–1.138, U0W–.457,
   (Arc Center = U.997W–.858), U.050W–1.767,
   (Arc Center = U.375W–.030) U.375W–.405, U.175W0,
   (Arc Center = U0W–.2) U.2W–.2, U0W–.350

Linear Interpolation Answers

Programming Exercise 3-1 Answer

The path tool path is programmed using linear interpolation (G01). Refer to Figure 3-2. The dashed line on the drawing indicates the net shape of the part and the metal to be removed.

Caution: DO NOT attempt to execute this program from solid bar stock.

O0001
(CNC Turning Center Programming Exercise 1)
(Date, By)
(Tool #4, Finish Turning Tool)
N10 G50 S2500
N15 T0400
N20 G96 S512 M03
N25 G00 G54 X2.35 Z.1 T0404 M08
N30 G1 Z0.0 F.0215
N35 X–.01
N40 Z.1
N45 G00 X.875
N50 G1 Z–.700
N55 X1.75
N60 Z–1.455
N65 X2.25
N70 G00 X2.35 Z.1 M09
N75 T0400
N80 G28 U0.0 W0.0
N85 M30

Programming Exercise 3-2 Answer

The tool path is programmed using linear interpolation (G01). The calculations necessary to offset the tool path for the Tool Nose Radius Compensation (TNRC) follow. The dashed line on the drawing indicates the net shape of the part and the metal to be removed. Refer to Figure 3-3.

Caution: DO NOT attempt to execute this program from solid bar stock.

A calculation is required to determine the taper angle.
875 – .4375 = .4375
.4375/2 = .2188
A = ARCTAN .2188/.379
Angle A = 30°

A calculation is required to offset the tool path for the tool nose radius on the start of the tapered portion.
a = b × TAN 30°
.031 × .5774 = .0179
X axis value = .4375 + (.0179 × 2) = .4733

A calculation is required to offset the tool path for the tool nose radius for the end of the tapered portion.

$a = b \times TAN\ 15°$
$.031 \times .2679 = .0083$
Z axis value $= .379 - .0083 = .3707$

A calculation is required to offset the tool path in the X-axis for the tool nose radius on the start of the first chamfer.

$a = b \times TAN\ 45°$
$.031 \times 1.0 = .031$
X axis value $= 1.438 + (.031 \times 2) = 1.5$

A calculation is required to offset the tool path in the Z-axis for the tool nose radius on the end of the first chamfer.

$a = b \times TAN\ 22.5°$
$.031 \times .4142 = .0128$
Z axis value $= .856 - .0128 = .8432$

A calculation is required to offset the tool path in the X-axis for the tool nose radius on the start of the second chamfer.

$a = b \times TAN\ 45°$
$.031 \times 1.0 = .031$
X axis value $= 2.09 + (.031 \times 2) = 2.152$

A calculation is required to offset the tool path in the Z-axis for the tool nose radius on the end of the second chamfer.

$a = b \times TAN\ 22.5°$
$.031 \times .4142 = .0128$
Z axis value $= 1.455 - .0128 = 1.4422$

```
O0002
(CNC Turning Center Programming Exercise 2)
(Date, By)
(Tool #4, Finish Turning Tool)
N10 G50 S2500
N15 T0400
N20 G96 S563 M03
N30 G00 G54 X.6975 Z.1 T0404 M08
N40 G1 Z0.0 F.022
N50 X-.01
```

N60 Z.1
N70 G00 X.4733
N80 G01 Z0
N90 X.875 Z–.3707
N100 Z–.7
N110 X1.5
N120 X1.75 Z–.8432
N130 Z–1.375
N140 X2.152
N150 X2.25 Z–1.4422
N160 Z–2.25
N170 G00 X2.35 Z.1 M09
N175 T0400
N180 G28 U0.0 W0.0
N190 M30

Programming Exercise 3-3 Answer

Using the Fixed Cutting Cycle B (G94), write a program to create the facing cut in Figure 3-4 in three equal depths of cut passes.

O0003
(CNC Turning Center Programming Exercise 3)
(Date, By)
(Tool #2, Rough Turning Tool)
N10 G50 S2500
N15 T0200
N20 G96 S513 M03
N30 G00 G54 X1.85 Z.1 T0202 M08
N40 G94 X–.03 Z–.0623 F.022
N50 Z–.1247
N60 Z–.187
N65 T0200 M09
N70 G28 U0.0 W0.0
N80 M30

Programming Exercise 3-4 Answer

Using the Fixed Cutting Cycle A (G90), write a program to create the turning cut in three equal depths of cut passes. Refer to Figure 3-5.

O0004
(CNC Turning Center Programming Exercise 4)
(Date, By)
(Tool #2, Rough Turning Tool)
N10 G50 S2500
N15 T0200
N20 G96 S513 M03
N30 G00 G54 X1.85 Z.1 T0202 M08
N40 G90 X1.625 Z–1.12 F.022
N50 X1.500
N60 X1.375
N65 T0200
N70 G28 U0.0 W0.0M09
N80 M30

Linear and Circular Interpolation Answers

Programming Exercise 3-5 Answer

The tool path is programmed using linear and circular interpolation (G01 and G03). Refer to Figure 3-6. The dashed line on the drawing indicates the net shape of the part and the metal to be removed.

Caution: DO NOT attempt to execute this program from solid bar stock.

O0005
(CNC Turning Center Programming Exercise 5)
(Date, By)
(Tool #4, Rough Turning Tool)
N10 G50 S6000
N20 T0400
N30 G96 S3650 M03
N40 G00 G54 X0.0 Z.1 T0404 M08
N50 G1 Z.031 F.0265
N60 X.187
N70 G3 X.937 Z–.344 R.406
N80 G1 Z–.669
N90 X1.125
N100 G3 X1.812 Z–.981 R.312
N110 G1 Z–1.344
N120 G3 X2.312 Z–1.656 R.312
N130 G1 Z–2.25
N140 G00 X2.35 Z.1

N145 T0400 M09
N150G28U0W0
N160 M30

The following is the same program written using the I and K commands for the arcs.

O0005
(CNC Turning Center Programming Exercise 5)
(Date, By)
(Tool #1, Rough Turning Tool)
N10 G50S6000
N20 T0400
N30 G96 S3650 M03
N40 G00 G54 X0.0 Z.1 T0404 M08
N50 G1 Z.031 F.0265
N60 X.187
N70 G3 X.937 Z–.344 I0.0 K–.406
N80 G1 Z–.669
N90 X1.25
N100 G3 X1.812 Z–.981 I0.0 K–.312
N110 G1 Z–1.344
N120 G3 X2.312 Z–1.656 I0.0 K–.312
N130 G1 Z–2.25
N140 G00 X2.35 Z.1
N145 T0400 M09
N150 G28 U0.0 W0.0
N160 M30

Programming Exercise 3-6 Answer

The tool path is programmed using linear and circular interpolation (G01, G02, and G03). Refer to Figure 3-7. The dashed line on the drawing indicates the net shape of the part and the metal to be removed.

Caution: DO NOT attempt to execute this program from solid bar stock.

O0006
(CNC Turning Center Programming Exercise 6)
(Date, By)
(Tool #4, Finish Turning Tool)
N10 G50 S6000
N20 T0400

```
N30 G96 S3650 M03
N40 G00 G54 X0.0 Z.1 T0404 M08
N50 G1 Z.031 F.0265
N60 G3 X.777 Z–.3575 R.3885
N70 G1 Z–.655
N80 G2 X.965 Z–.749 R.094
N90 G1 X1.252
N100 G3 X1.688 Z–.967 R.218
N110 G1 Z–1.255
N120 G2 X2.066 Z–1.444 R.189
N130 G1 X2.35
N140 G3 X2.512 Z–1.525 R.081
N150 G1 Z–2.25
N160 G00 X2.55 Z.1 M09
N165 T0404
N170 G28 U0.0 W0.0
N180 M30
```

The following is the same program written using the I and K commands for the arcs.

```
O0006
(CNC Turning Center Programming Exercise 6)
(Date, By)
(Tool #4, Finish Turning Tool)
N10 G50 S6000
N20 T0400
N30 G96 S3650 M03
N40 G00 X0.0 Z.1 T0404 M08
N50 G1 Z.031 F.0265
N60 G3 X.777 Z–.3575 I0.0 K–.3885
N70 G1 Z–.655
N80 G2 X.965 Z–.749 I.094 K0.0
N90 G1 X1.252
N100 G3 X1.688 Z–.967 I0.0 K–.218
N110 G1 Z–1.255
N120 G2 X2.066 Z–1.444 I.189 K0.0
N130 G1 X2.35
N140 G3 X2.512 Z–1.525 I0.0 K–.081
N150 G1 Z–2.25
N160 G00 X2.55 Z.1
N165 T0400 M09
```

N170 G28 U0.0 W0.0
N180 M30

Tool Nose Radius Compensation Answers

Note: The correct values must be set in the offset register for each of the tools, identifying the amount of the tool nose radius compensation. Even if they are called properly in the program, the correct result will not occur if these values are not set properly. This method will differ, to some degree, dependent on the specific controller used. You should consult the Operation and Programming manuals specific to the machine and controller being used.

Programming Exercise 3-7 Answer

The following program is the toolpath using linear and circular interpolation and TNRC (G40, G41 and G42) for the toolpath contours in Programming Exercise 3-2. Refer to Figure 3-3. The dashed line on the drawing indicates the net shape of the part and the metal to be removed.

Caution: DO NOT attempt to execute this program from solid bar stock.

```
O0007
(CNC Turning Center Programming Exercise 7)
(Date, By)
(Tool #4, Finish Turning Tool)
N10 G50 S3500
N15 T0400
N20 G96 S563 M03
N30 G00 G54 G41 X.7 Z.1 T0404 M08
N40 G1 Z0.0 F.022
N50 X-.01
N60 G00 Z.1
N70 G42 X.4375
N80 G1 Z0.0
N90 X.875 Z-.379
N100 Z-.7
N110 X1.438
N120 X1.75 Z-.856
N130 Z-1.375
N140 X2.09
N150 X2.25 Z-1.455
N160 Z-2.25
N170 G00 G40 X2.35 Z.1
```

N175 T0400 M09
N180 G28 U0.0 W0.0
N190 M30

Programming Exercise 3-8 Answer

Refer to Figure 3-6 and Programming Exercise 3-8. The dashed line on the drawing indicates the net shape of the part and the metal to be removed.

Caution: DO NOT attempt to execute this program from solid bar stock.

```
O0008
(CNC Turning Center Programming Exercise 8)
(Date, By)
(Tool #1, Finish Turning Tool)
N10 G50 S6000
N20 T0400
N30 G96 S3650 M03
N40 G00 G54 G42 X0.0 Z.1 T0404 M08
N50 G1 Z0.0 F.0265
N60 X.125
N70 G3 X.875 Z–.375 R.375
N80 G1 Z–.7
N90 X1.188
N100 G3 X1.75 Z–.981 R.281
N110 G1 Z–1.375
N120 G3 X2.25 Z–1.656 R.281
N130 G1 Z–2.25
N140 G00 G40 X2.35 Z.1 M09
N145 T0400
N150 G28 U0.0 W0.0
N160 M30
```

The following is the same program written using the I and K commands for the arcs.

```
O0008
(CNC Turning Center Programming Exercise 8)
(Date, By)
(Tool #4, Finish Turning Tool)
N10 G50 S6000
N20 T0400
N30 G96 S3650 M03
N40 G00 G54 G42 X0.0 Z.1 T0404 M08
N50 G1 Z0.0 F.0265
N60 X.125
```

```
N70 G3 X.875 Z-.375 I0.0 K-.375
N80 G1 Z-.7
N90 X1.188
N100 G3 X1.75 Z-.981 I0.0 K-.281
N110 G1 Z-1.375
N120 G3 X2.25 Z-1.656 I0.0 K-.281
N130 G1 Z-2.25
N140 G00 G40 X2.35 Z.1
N145 T0400 M09
N150 G28 U0.0 W0.0
N160 M30
```

Programming Exercise 9 Answer

Refer to Figure 3-7 and Exercise 3-6. The dashed line on the drawing indicates the net shape of the part and the metal to be removed.

Caution: DO NOT attempt to execute this program from solid bar stock.

```
O0009
(CNC Turning Center Programming Exercise 9)
(Date, By)
(Tool #4, Finish Turning Tool)
N10 G50 S6000
N20 T0400
N30 G96 S3650 M03
N40 G00 G54 G42 X0.0 Z.1 T0404 M08
N50 G1 Z0.0 F.0265
N60 G3 X.715 Z-.3575 R.3575
N70 G1 Z-.655
N80 G2 X.965 Z-.780 R.125
N90 G1 X1.252
N100 G3 X1.626 Z-.967 R.187
N110 G1 Z-1.255
N120 G2 X2.066 Z-1.475 R.22
N130 G1 X2.35
N140 G3 X2.45 Z-1.525 R.05
N150 G1 Z-2.25
N160 G00 G40 X2.55 Z.1
N165 T0400 M09
N170 G28 U0.0 W0.0
N180 M30
```

The following is the same program written using the I and K commands for the arcs.

```
O0009
(CNC Turning Center Programming Exercise 9)
(Date, By)
(Tool #4, Finish Turning Tool)
N10 G50 S6000
N20 T0400
N30 G96 S3650 M03
N40 G00 G54 G42 X0.0 Z.1 T0404 M08
N50 G1 Z0.0 F.0265
N60 G3 X.715 Z–.3575 I0.0 K–.3575
N70 G1 Z–.655
N80 G2 X.965 Z–.780 I.125 K0.0
N90 G1 X1.252
N100 G3 X1.626 Z–.967 I0.0 K–.187
N110 G1 Z–1.255
N120 G2 X2.066 Z–1.475 I.22 K0.0
N130 G1 X2.35
N140 G3 X2.45 Z–1.525 I0.0 K–.05
N150 G1 Z-2.25
N160 G00 G40 X2.55 Z.1
N165 T0400 M09
N170 G28 U0.0 W0.0
N180 M30
```

Drilling Answers

Programming Exercise 3-10 Answer

The tool path is programmed using linear interpolation (G01) to center drill the face of the part (Figure A3-1).

Calculate the depth of cut required using a #5 (Plain Type) center drill to countersink to a diameter of .405 inch. Plain Type center drills have an angle of 60° with a point angle of 120°. The length from the end of the point angle to the beginning of the 60° angle is 3/16 inch (See Center Drill table in *Machinery's Handbook*, 29th Edition).

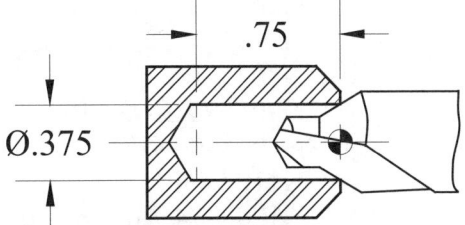

Figure A3-1   CNC Turning Center Programming Exercise 3-10

.0938 × TAN 30° = .054 (for the 120° tip)
.109 × TAN 60° = .1888 (for the 60° portion)
.054 + .1888 + .1875 = .4303

O0010
(CNC Turning Center Programming Exercise 10)
(Date, By)
(Tool 1 = #5 Center Drill)
N10 T0100
N20 G97 S641 M03
N30G00 G54 X0.0 Z.1 T0101 M08
N40 G1 Z–.4303 F.009
N50 G00 Z.1 M09
N55 T0100
N60 G28 U0.0 W0.0
N70 M30

Programming Exercise 3-11 Answer

G74

The tool path is programmed using the drilling cycle G74 with three equal depth cuts. Refer to Figure 3-9.

O0011
(CNC Turning Center Programming Exercise 11)
(Date, By)
(Tool 1 = #5 Center Drill)
N10 T0100
N20 G97 S641 M03
N30 G00 G54 X0.0 Z.1 T0101 M08
N40 G1 Z-.4303 F.009
N50 G00 Z.1 M09
N60 G28 U0.0 W0.0
N65 T0100
N70 M01
(Tool 7 = .375 Diameter HSS Drill)
N80T0700
N90 G97 S693 M03
N100 G00 G54 X0.0 Z.1 T0707 M08
N110 G74 R.1
N120 G74 X0.0 Z–1.0 Q.333 F.009
N130 G00 Z.1 M09
N140 G28 U0.0 W0.0
N145 T0700
N150 M30

Multiple Repetitive Cycles

Programming Exercise 3-12 Answer

G71 and G70 Rough and Finish Turn Cycle

The tool path is programmed using Turning Cycles (G71 and G70). Refer to Figure 3-10.

```
O0012
(CNC Turning Center Programming Exercise 12)
(Date, By)
(Tool #2, Rough Turning Tool)
N10 G50 S6000
N20 T0200
N30 G96 S2800 M3
N40 G00 G54 G41 X3.1 Z.1 T0202 M08
N50 G1 Z0.0 F.0265
N60 X–.01
N70 G00 Z.1
N80 G42 X3.0
N90 G71 U.08 R.1
N100 G71 P110 Q210 U.015 W.005
N110 G00 X.94
N115 G01 Z0.0 F.0265
N120 X1.0 Z–.03
N130 Z–.94
N140 G2 X1.12 Z–1.0 R.06
N150 G01 X1.8
N160 X2.0 Z–1.1
N170 Z–1.94
N180 G2 X2.12 Z–2.0 R.06
N190 G01 X2.8
N200 X3.0 Z–2.1
N210 X3.1
N215 T0200
N220 G28 G40 U0.0 W0.0
N230 M01
(Tool #4, Finish Turning Tool)
N240 G50 S6000
N250 T0400
N260 G96 S2800 M3
N270 G00 G54 G42 X3.1 Z.1 T0404 M08
N280 G70 P110 Q210
N285 T0400
N290 G28 G40 U0.0 W0.0 M09
N300M30
```

Note: To use the single block format for program #O0012, omit line N90 and replace line N100 with: N100 G71 P110 Q210 U.015 W.005 D.08

Boring Answers

Programming Exercise 3-13 Answer

G71 and G70 Rough and Finish Turn Cycle

The same type of tool path is programmed using Turning Cycles (G71 and G70) and applied to internal boring operations. Refer to Figure 3-11.

```
O0013
(CNC Turning Center Programming Exercise 13)
(Date, By)
(Tool #3, Rough Boring Tool)
N10 G50 S6000
N20 T0300
N30 G96 S2800 M3
N40 G00 G54 G41 X.713 Z.1 T0303 M08
N50 G71 U.08 R.03
N60 G71 P70 Q140 U.015 W.005
N70 G00 X2.192
N80 G01 Z0.0 F.0265
N90 G2 X2.132 Z–.03 R.03
N100 G01 Z–.482
N110 X1.714 Z–.825
N120 X1.209
N130 X.813 Z–1.025
N140 X.713
N145 T0300
N150 G28 G40 U0.0 W0.0
N160 M01
(Tool #5, Finish Boring Tool)
N170 G50 S6000
N180 T0500
N190 G96 S2800 M3
N200 G00 G42 X.90 Z.1 T0505 M08
N210 G70 P80 Q140
N215 T0400 M09
N220 G28 G40 U0.0 W0.0
N230 M30
```

Note: To use the single block format for program #O0013, omit line N50 and replace line N60 with: N60 G71 P70 Q140 U.015 W.005 D.08

Programming Exercise 3-14 Answer

G72 Face Cutting Cycle

The tool path is programmed using the Face Cutting Cycle (G72). Refer to Figure 3-12.

```
O0014
(CNC Turning Center Programming Exercise 14)
(Date, By)
(Tool #2, Rough Turning Tool)
N10 G50 S6000
N20 T0200
N30 G96 S2800 M3
N40 G00 G54 G42 X2.005 Z.1 T0202 M08
N50 G72 U.06 R.1
N60 G72 P70 Q190 U.015 W.015
N70 G00 Z0.0 F.0265
N80 X.28
N90 G3 X.4 Z–.06 R.06
N100 G01 Z–.125 F.0265
N110 X.8 Z–.325
N120 X1.185
N130 G3 X1.305 Z–.385 R.06
N140 G01 Z–.433
N150 G2 X1.365 Z–.463 R.03
N160 G1 X1.785
N170 G3 X1.905 Z–.523 R.06
N180 G01 Z–.563
N190 X2.005
N195 T0100 M09
N200 G28 G40 U0.0 W0.0
N210 M01
(Tool #4, Finish Turning Tool)
N220 G50 S6000
N230 T0400
N240 G96 S2800 M3
N250 G00 G54 G42 X.90 Z.1 T0404 M08
N260 G70 P80 Q190
N265 T0400 M09
N270 G28 G40 U0.0 W0.0
N280 M30
```

Note: To use the single block format for program #O0014, omit line N50 and replace line N60 with: N100 G72 P70 Q190 U.015 W.005 D.06

Programming Exercise 15 Answer

G73 Pattern Repeating Cycle

The tool path is programmed using Pattern Repeating Cycle (G73). Refer to Figure 3-13. The dashed line on the drawing indicates the net shape of the part and the metal to be removed.

Caution: DO NOT attempt to execute this program from solid bar stock.

A calculation is required to obtain the length of the tapered section.

$$.720 - .409 = .311$$
$$.311/2 = .1555$$
$$\text{Length} = .155 \times \text{TAN } 75° = .5803$$

```
O0015
(CNC Turning Center Programming Exercise 15)
(Date, By)
(Tool #2, Rough Turning Tool)
N10 G96 S513 M3
N20 T0200
N30 G00 G54 G42 X2.338 Z.1 T0202 M08
N40 G1 Z.05 F.0215
N50 G73 U.2 W.2 R3
N60 G73 P70 Q160 U.03 W.005
N70 G00 X.349
N80 G1 Z0.0 F.0215
N90 G3 X.409 Z–.03 R.03
N100 G1 Z–.526
N110 X.72 Z–1.1063
N120 Z–1.662
N130 X1.574 Z–2.089
N140 X1.778
N150 G3 X1.838 Z–2.119 R.03
N160 G1 X2.338
N165 T0200
N170 G28 G40 U0.0 W0.0
N180 M01
(Tool #4, Finish Turning Tool)
N190 G96 S513 M3
```

N200 T0400
N210 G00 G54 G42 X2.338 Z.1 T0404 M08
N220 G1 Z.05 F.0215
N230 G70 P70 Q160
N235 T0400 M09
N240 G28 G40 U0.0 W0.0
N250 M30

Note: To use the single block format for program #O0015, omit line N50 and replace line N60 with: N100 G73 P70 Q160 U.03 W.005 D3

Programming Exercise 16 Answer

The same Pattern Repeating Cycle type is applied to internal boring using (G73). Refer to Figure 3-14. The dashed line on the drawing indicates the net shape of the part and the metal to be removed. The .813 diameter has been predrilled in an earlier operation.

Caution: DO NOT attempt to execute this program from solid bar stock.

O0016
(CNC Turning Center Programming Exercise 16)
(Date, By)
(Tool #3, Rough Boring Tool)
N10 T0300
N20 G96 S513 M3
N30 G00 G54 G41 X.713 Z.1 T0303 M08
N40 G1 Z.05 F.0215
N50 G73 U.2 W.2 R3
N60 G73 P70 Q180 U.03 W.005
N70 G00 X2.869
N80 G1 Z0.0 F.0215
N90 G2 X2.809 Z–.03 R.03
N100 G1 Z–.19
N110 G3 X2.689 Z–.25 R.06
N120 G1 X2.475
N130 G2 X2.532 Z–.28 R.03
N140 G1 Z–.625
N150 X1.46 Z–1.075
N160 X1.063
N170 G2 X.813 Z–1.2 R.125
N180 G1 X.713
N185 T0100 M09
N190 G28 G40 U0.0 W0.0
N200 M01
(Tool #5, Finish Boring Tool)

N210 T0500
N220 G96 S513 M3
N230 G00 G54 G42 X2.338 Z.1 T0505 M08
N240 G1 Z.05 F.0215
N250 G70 P70 Q180
N255 T0200 M09
N260 G28 G40 U0.0 W0.0
N270 M30

Note: To use the single block format for program #O0016, omit line N50 and replace line N60 with: N100 G73 P70 Q180 U.03 W.005 D3

Grooving Answers

Programming Exercise 3-17 Answer

G75

The tool path is programmed using Groove Cutting Cycle (G75). Use Figure 3-15. The major diameter is prepared prior to this operation.

O0017
(CNC Turning Center Programming Exercise 17)
(Date, By)
(Tool #6, OD Grooving Tool .118 Wide)
N10 T0600
N20 G96 S750 M3
N30 G00 G54 X1.53 Z.1 T0606 M08
N40 Z–.912
N50 G75 R.1
N60 G75 X1.23 Z–.912 K.1
N70 G00 X1.53
N75 T0600 M09
N80 G28 G40 U0.0 W0.0
N90 M30

OD Threading Answers

Programming Exercise 3-18 Answer

G76 Threading Cycle

The tool path is programmed using Threading Cycle (G76). Refer to Figure 3-16. The major diameter is prepared prior to this operation.

O0018
(CNC Turning Center Programming Exercise 18)
(Date, By)
(Tool #8, OD Threading Tool)
N10 T0800
N20 G97 S764 M3
N30 G00 G54 X1.1 Z.1 T0808 M08
N40 G76 X.8492 Z–.93 K.0707 D0150 A60 F.125
N45 T0800 M09
N50 G28 G40 U0.0 W0.0
N60 M30

CNC Turning Center Subprogram Application Answers

Programming Exercise 3-19 Answer

M98 and M99

The tool path is programmed using linear and circular interpolation with the application of a subprogram call and return (M98 and M99). Refer to Figure 3-17. The outside diameter is prepared in a prior operation.

O0019
(CNC Turning Center Programming Exercise 19)
(Date, By)
(Tool #6, OD Grooving Tool .118 Wide)
N10 G50 S6000
N20 T0600
N30 G96 S2100 M3
N40 G00 G54 X1.9 Z.1 T0606 M08
N50 Z–.407
N60 M98 P3456 L1
N70 G00 Z-.934
N80 M98 P3456 L1
N90 G00 Z–1.496
N100 M98 P3456 L1
N115 T0500 M09
N110 G28 G40 U0.0 W0.0
N120 M30

Subprogram for Program O0019
O3456

N1 G01 X1.4 F.0265
N2 G00 X1.9
N3 Z–.256
N4 G1 X1.8 F.013
N5 U–.05 W–.05
N6 U–.105
N7 G2 U–.045 W–.045 R.045
N8 G00 X1.9
N9 G1 Z–.488
N10 X1.8
N11 U–.05 W.05
N12 U–.105
N13 G3 U–.045 W.045 R.045
N14 G00 X1.9
M9

CNC Turning Center Combined Projects Answers

Programming Exercise 3-20 Answer

The tool path is programmed using several Turning Cycles. Refer to Chart A3-1.

O0020
(CNC Turning Center Program-
  ming Exercise 20)
(Date, By)
(Tool #2, Rough Turning Tool)
N10 G50 S6000
N20 T0200
N30 G96 S563 M3
N40 G00 G54 G42 X2.6 Z.1
  T0202 M08
N50 G1 Z0.0 F.022
N60 X–.02
N70 G00 Z.1
N80 G42 X.2.6
N90 G71 U.08 R.1
N100 G71 P110 Q210 U.015
  W.005
N110 G00 X.90
N120 G01 Z0.0 F.0265
N130 X1.0 Z–.05
N140 Z–1.0

N150 X1.3
N160 G03 X1.5 Z–1.1 R.1
N170 G01 Z–1.9
N180 G2 X1.7 Z–2.0 R.1
N190 G01 X2.0 Z–2.0
N200 X2.5 Z–2.25
N210 X2.6
N215 T0200 M09
N220 G28 G40 U0.0 W0.0
N230 M01
(Tool #4, Finish Turning Tool)
N240 G50 S6000
N250 T0400
N260 G96 S563 M3
N270 G00 G54 G42 X.90 Z.1
  T0404 M08
N280 G1 Z0.0 F.022
N290 G70 P110 Q210
N295 T0400 M09
N300 G28 G40 U0.0 W0.0
N310 M01

Turning Center
CNC Setup Sheet

Date: Today	**Prepared By:** You
Part Name: Turning Center Project	**Part Number:** 1234
Machine: CNC Turning Center	**Program Number:** O0020

Workpiece Zero: **X** _Centerline_   **Y** _NA_   **Z** _Finished Face_

Setup Description:

Clamp the part in a 3-jaw chuck with soft-jaws, 2.35" minimum extended out of the chuck.

Tool Number	Offset Number	Tool Description	Comments
02	02	Rough Turning Tool .031 TNR	SFPM 125–1000
04	04	Finish Turning Tool .015 TNR	SFPM 125–1000
06	06	OD Grooving Tool	.118 Wide .005 TNR
08	08	OD Threading Tool	SFPM 125–1000

Facing Cut = r/min 191–1527
Turning 1.0" diameter, r/min = 477–3820
Turning 1.5" diameter, r/min = 318–2546
Turning 1.0" diameter, r/min = 191–1528
Turning 1.0" diameter, r/min = 582–4658
Threading r/min = 477–3820 Feed = Thread Lead
NOTE: The preferred method of setting the r/min for threading is to use the Constant Cutting Speed (G96) command.

(Tool #6, OD Grooving Tool
.118 Wide)
N320 T0600
N330 G96 S563 M3
N340 G00 X1.6 Z–.906 T0606
M08
N350 X1.1
N360 G75 R.1

N370 G75 X.82 Z–.906 I.09
N380 G00 X2.5
N385 T0600 M09
N390 G28 G40 U0.0 W0.0
N400 M01
(Tool #7, OD Threading Tool)
N410 T0800
N420 G97 S1074 M3

N430 G00 G54 X1.1 Z.1 T0808
  M08
N440 G76 X.8492 Z–.8 K.0707
  D0150 A60 F.125

N445 T0800 M09
N450 G28 G40 U0.0 W0.0
N460 M30

Note: To use the single block format for program #O0020, omit line N90 and replace line N100 with: N100 G71 P110 Q210 U.015 W.005 D.08

CNC Turning Center Program Error Diagnosis Answers

Use the skills you have learned to identify the problems in the following program lines and program sections. You may refer to the text, *Programming of CNC Machines*, Fourth Edition.

1. Use the following CNC code and sketch a representation of the part being created. Refer to Figure A3-2.

O2001
(CNC Turning Center Program Diagnosis, Problem 1)
(Tool #2, Rough Turning Tool)
N10 G50 S2000
N15 T0200 M42
N20 G96 S500 M03
N25 G00 G54 X2.2 Z.3 T0202 M08
N30 G01 Z.01 F.03
N35 X0.0 F.012
N40 G00 X3.0 Z.2
N45 G73 P50 Q85 I.168 K.169 U.04
  W.02 D3 F.012
N50 G00 X1.59
N55 G01 Z0.0
N60 X1.75 Z–.08
N65 Z–1.375
N70 X2.0 W–.125
N75 Z–2.1
N80 G03 U.3 Z–2.25 I–.15 K0.0 F.004
N85 G01 X2.75
N90 Z–3.75
N95 X2.85
N96 T0100 M09
N100 G28 U0.0 W0.0
N105 M30

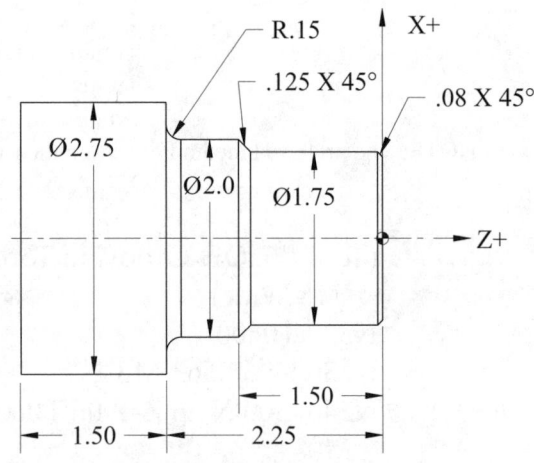

Figure A3-2   CNC Turning Center Programming Error Diagnosis Part Drawing

2. Find the error in this program line.

In line N100, the arc center coordinates or R for radius is missing.

> N100 G03 U.3 Z-2.25 **I–.15 K0.0** F.004

or

> N100 G03 U.3 Z–2.25 **R.15** F.004

3. Find the error in the following program lines.

In line N40, the federate is missing for the G01 function.

> O2003
> (CNC Turning Center Program Diagnosis, Problem 3)
> (Tool 1 #5 Center Drill)
> N10 T0100
> N20 G97 S641 M03
> N30 G00 G54 X0.0 Z.1 T0101 M08
> N40 G1 Z–.4303 **F.006**
> N50 G0.0 Z.1 M09
> N60 G28 U0.0 W0.0 T0100
> N70 M30

4. Find the error in the following program section.

In line N20, the spindle direction command is missing.

> O2005
> (CNC Turning Center Program Diagnosis, Problem 5)
> (Tool #2, Rough Turning Tool)
> N10T0200
> N20 G50 S6000 **M03**
> N30 G96 S3650
> N40 G00 G54 X0.0 Z.1 T0202 M08
> . . . .
> . . . .

5. Find the error in the following program.

In line N60, the improper line numbers are called to execute the Multiple Repetitive Cycle.

> O2005
> (CNC Turning Center Program Diagnosis, Problem 5)
> (Tool #3, Rough Boring Tool)
> N10 T0300
> N20 G50 S6000
> N30 G96 S2800 M3
> N40 G00 G54 G41 X.713 Z.1 T0303 M08
> N50 G71 U.08 R.03
> N60 G71 **P50 Q150** U.015 W.005
> N70 G0 X2.192
> N80 G01 Z0.0 F.0265
> N90 G2 X2.132 Z–.03 R.03
> N100 G01 Z–.482
> N110 X1.714 Z–.825
> N120 X1.209
> N130 X.813 Z–1.025
> N140 X.713
> N145 T0300 M09
> N150 G28 G40 U0.0 W0.0
> N160 M01

6. Find the error in the following program section.

The program is missing a tool nose radius compensation call (T0202).

> O2007
> (CNC Turning Center Program Diagnosis, Problem 7)
> (Tool #2, Rough Turning Tool)
> N10 T0200
> N20 G96 S563 M03
> N30 G00 G41 X.6975 Z.1 **T0101** M08
> N40 G1 Z0.0 F.022
> N50 X-.01
> N60 G00 Z.1
> N70 G42 X.4375
> . . . .
> . . . .

7. Find the error in the following program.

   In line N70, the tool nose radius compensation direction is incorrect. It should be G42.

   O2008
   (CNC Turning Center Program Diagnosis, Problem 7)
   (Tool #2, Rough Turning Tool)
   N10 T0200
   N20 G96 S563 M03
   N25 T0202 M08
   N30 G00 G54 G41 X.6975 Z.1
   N40 G1 Z0.0 F.022
   N50 X–.01
   N60 G00 Z.1
   N70 **G42** X.4375
   N80 G1 Z0.0
   N90 X.875 Z–.379
   N100 Z–.7
   N110 X1.438
   N120 X1.75 Z–.856
   N130 Z–1.375
   N140 X2.09
   N150 X2.25 Z–1.455
   N160 Z–2.25
   N170 G00 X2.35 Z.1 M09
   N180 G28 G40 U0.0 W0.0
   N190 M30

8. Find the error in the following program section.

   In line N100, re-initiation of G01 linear interpolation is missing.

   O2009
   (CNC Turning Center Program Diagnosis, Problem 8)
   (Tool #2, Rough Turning Tool)
   N10 T0200
   N20 G50 S6000
   N30 G96 S2800 M3
   N35 T0202 M08
   N40 G00 G54 G42 X2.005 Z.1
   N50 G72 U.06 R.1
   N60 G72 P70 Q190 U.015 W.015
   N70 G00 Z0.0 F.0265

N80 X.28
N90 G3 X.4 Z–.06 R.06
N100 **G01** Z–.125 F.0265
N110 X.8 Z–.325
N120 X1.185
N130 G3 X1.305 Z–.385 R.06
N140 G01 Z–.433
N150 G2 X1.365 Z–.463 R.03
N160 G1 X1.785
N170 G3 X1.905 Z–.523 R.06
N180 G01 Z–.563
N190 X2.005
N195 T0200 M09
N200 G28 G40 U0.0 W0.0
N210 M01

9. Find the error in the following program section.

In line N240, the cutting speed is missing from the Constant Surface Speed call.

(Tool #4, Finish Turning Tool)
N220 T0400
N230 G50 S6000
N240 G96 M3
N250 G00 G54 G42 X.90 Z.1 T0404 M08
N260 G70 P80 Q190
N265 T0400 M09
N270 G28 G40 U0.0 W0.0
N280 M30

10. Find the error in the following program section.

At the end of the finishing cycle, the cutter compensation cancellation is missing.

(Tool #5, Finish Boring Tool)
N170 T0500
N180 G50 S6000
N190 G96 S2800 M3
N200 G00 G54 G42 X.90 Z.1 T0505 M08
N210 G70 P80 Q140
N 215 **T0400** M09
N220 G28 U0.0 W0.0
N230 M30

Answer Key

Answers to Unit 4: CNC Machining Center Programming

The answer programs given here are one method of machining the parts shown in the drawings. There are many possibilities for getting the correct results. The student should consult with their instructor for program verification and always proceed, with caution, when running a newly written program. In the following exercises where only one tool is required, a CNC Setup sheet is not necessary but the tool and setup information should be listed before your program code. Please use a CNC Setup sheet in all other cases for the sake of clarity.

CNC Machining Center Tool List

Face Mill, 3.0 inch diameter, 90°, 5 teeth, Carbide

End Mill, 2-Flute, 1/8 inch
End Mill, 2-Flute, 3/8 inch
End Mill, 2-Flute, 9/16 inch
End Mill, 2-Flute, 1 inch

End Mill, 4-Flute, 3/8 inch
End Mill, 4-Flute, 1/2 inch
End Mill, 4-Flute, 5/8 inch
End Mill, 4-Flute, 3/4 inch
End Mill, 4-Flute, 1 inch

Roughing End Mill, 4-Flute, 1 inch

#5 Center Drill
#6 Center Drill

Spotting Drill, .75 diameter, 90° single flute

Drill Bits, All sets are available in High Speed Steel

Taps, All sizes are available in High Speed Steel

Reamers, all required sizes are available in High Speed Steel

Identifying Programming Coordinates for Milling Answers

1. Absolute programming coordinates, including the arc center locations. Refer to Figure 4-1.

   X0Y0, X0Y1.0, X2.0Y1.75 (Arc Center X3.0Y1.75) X4.0Y1.75,
   (Arc Center X4.0Y1.375) X4.0Y1.0, X4.0Y0, X3.25Y0,
   X3.25Y.313 (Arc Center X3.063Y.313) X3.063Y.5, X.687Y.5,
   (Arc Center X.687Y.313) X.5Y.313, X.5Y0, X0Y0

2. Incremental programming coordinates, including the arc center locations. Refer to Figure 4-1.

   X0Y0, X0Y1.0, X2.0Y.75 (Arc Center X1.0Y0) X2.0Y0,
   (Arc Center X0Y–.375) X0Y–75, X0Y–1.0, X–.75Y0,
   X0Y.5 (Arc Center X–.187Y.313) X–2.75Y0,
   (Arc Center X.687Y–.187) X0Y–.5, X–.5Y0

Linear Interpolation on the CNC Machining Center

Face Milling Answers

Programming Exercise 4-1 Answer

The tool path is programmed using linear interpolation and rapid positioning (G01 and G00) for the Machining Center. Refer to Figure 4-2.

Tool = Face Mill, 3.0 inch diameter, 90°, 5 teeth, Carbide
Cutting Speed = 39–475
r/min = 50–605
in/tooth = .020–.039
in/min = 5.0–118

```
O0001
(CNC Machining Center Exercise #1)
(Date, by)
G90 G20 G80 G40 G49
(Tool #1 Face Mill, 3.0 inch diameter, 90°, 5 teeth, Carbide)
N20 T01 M06
N30 S328 M03
N40 G00 G54 X5.85 Y–.65
N50 G43 Z1.0 H01
N60 Z.1 M08
N70 Z–.08 F10.0
N80 X–1.6 F49.0
N90 G80 Z.1 M09
N100 G91 G28 Z0.0
N110 G28 X0.0 Y0.0
N120 M30
```

Midrange feeds and speeds are used. On line N40, the Y-axis is positioned at –.5 in order that the tool is not positioned exactly on centerline. This is a better cutting method than positioning the tool on centerline. Consult the insert manufacturer catalogs to verify this technique.

Programming Exercise 4-2 Answer

The tool path is programmed using linear interpolation and rapid positioning (G01 and G00) for the Machining Center. Refer to Figure 4-3

Tool = Face Mill, 3.0 inch diameter, 90°, 5 teeth, Carbide
Cutting Speed = 755–1720
r/min = 961–2190
in/tooth = .020–.039
in/min = 96.0–427.0

```
O0002
(CNC Machining Center Exercise #2)
(Date, by)
N10 G90 G20 G80 G40 G49
(Tool #1 Face Mill, 3.0 inch diameter, 90°, 5 teeth, Carbide)
N20 T01 M06
N30 S1575 M03
N40 G00 G54 X5.85 Y–.5
N50 G43 Z1.0 H01
N60 Z.1 M08
N70 Z–.07 F10.0
N80 X–1.6 F236.0
N90 G00 Z.1
N100 X5.85 Y–3.0
N110 G01 Z–.07 F10.0
N120 X–1.6 F236.0
N90 G80 Z.1 M09
N100 G91 G28 Z0.0
N110 G28 X0.0 Y0.0
N120 M30
```

Midrange feeds and speeds are used.

Contour Milling Answers

Programming Exercise 4-3 Answer

The tool path for the part contour is programmed using linear interpolation and rapid positioning (G01 and G00) for the Machining Center. Refer to Figure 4-4.

Tool = .375 inch diameter, 4 flute, High Speed Steel, End Mill
Cutting Speed = 5–85
r/min = 51–865
in/tooth = .001–.004
in/min = .20–14.0

O0003
(CNC Machining Center Exercise #3)
(Date, By)
N10 G90 G20 G80 G40 G49
(Tool #1 .375 inch diameter, 4-flute High Speed Steel, End Mill)
N20 T01 M06
N30 S458 M03
N40 G54 G00 X–.287 Y0.0
N50 G43 Z1.0 H01
N60 Z.1 M08
N70 Z–.125 F10.0
N80 X6.031 F7.0
N90 Y–3.031
N100 X0.0
N110 Y.287
N120 G00 Z.1
N130 X–.287 Y0.0
N140 Z–.25 F10.0
N150 X6.031 F7.0
N160 Y–3.031
N170 X0.0
N180 Y.287
N190 G80 Z.1 M09
N200 G91 G28 Z0.0
N210 G28 X0.0 Y0.0
N220 M30

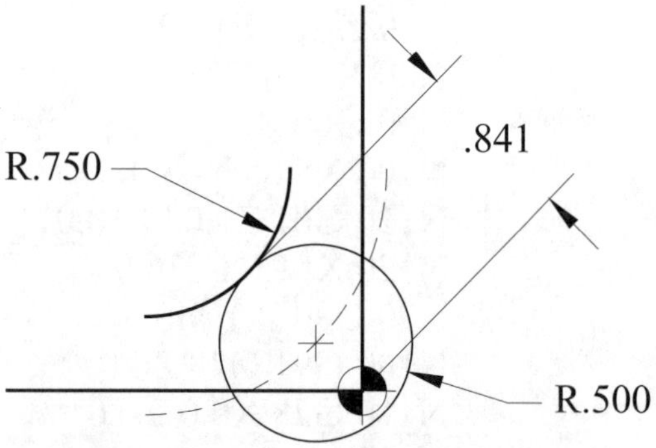

Figure A4-1   CNC Machining Center
Programming Exercise 4-4, Cutter Diamond
Calculation

Circular Interpolation Answers

Programming Exercise 4-4 Answer

The tool path is programmed for the part contour using linear and circular interpolation and rapid positioning (G01, G02, and G00), for the Machining Center. Radius programming is given using the program words R, I, and J. Refer to Figure 4-5.

Figure A4-1 shows that the 1.0 diameter end mill is necessary to accomplish the axial cut in one step.

Tool = 1.00 inch diameter, 4 flute, High Speed Steel, End Mill
Cutting Speed = 5–85
r/min = 19–324
in/tooth = .001–.004
in/min = .076–5.2

The following program is written entirely using the R command for the radii.

```
O0004
(CNC Machining Center Exercise #4)
(Date, By)
N10 G90 G20 G80 G40 G49
(Tool #1 1.0 inch diameter, 4 flute, High Speed Steel, End Mill)
N20 T01 M06
N30 S172 M03
N40 G54 G00 X–4.625 Y–.6
N50 G43 Z1.0 H01
N60 Z.1 M08
N70 G01 Z–.1875 F10.0
N80 Y3.25 F2.6
N90 G2 X–3.875 Y4.0 R.75
N100 G1 X–.937
N110 G2 X.125 Y2.938 R1.062
N120 G1 Y1.125
N130 G2 X–1.125 Y–.125 R1.25
N140 G1 X–3.8125
N150 G2 X–4.625 Y.6875 R.8125
N160 G1 Z–.375
N170 Y3.25
N180 G2 X–3.875 Y4.0 R.75
N190 G1 X–.937
N200 G2 X.125 Y2.938 R1.062
N210 G1 Y1.125
N220 G2 X–1.125 Y–.125 R1.25
N230 G1 X–3.8125
N240 G2 X-4.625 Y.6875 R.8125
N250 G80 G00 Z.1 M09
N260 G91 G28 Z0.0
N270 G28 X0.0 Y0.0
N280 M30
```

The following program is written entirely using the I and J commands for the radii.

```
O0004
(CNC Machining Center Exercise #4)
(Date, By)
N10 G90G20G80G40G49
(Tool #1 1.0 inch diameter, 4-flute High Speed Steel, End Mill)
N20 T01 M06
N30 S172 M03
N40 G54 G00 X–4.625 Y–.6
N50 G43 Z1.0 H01
N60 Z.1 M08
N70 G01 Z–.1875 F10.0
N80 Y3.25 F2.6
N90 G2 X–3.875 Y4.0 I.75 J0.0
N100 G1 X–.937
N110 G2 X.125 Y2.938 I0.0 J–1.062
N120 G1 Y1.125
N130 G2 X–1.125 Y–.125 I–1.25 J0.0
N140 G1 X–3.8125
N150 G2 X–4.625 Y.6875 I0.0 J.813
N160 G1 Z–.375
N170 Y3.25
N180 G2 X–3.875 Y4.0 I.75 J0.0
N190 G1 X–.937
N200 G2 X.125 Y2.938 I0.0 J–1.062
N210 G1 Y1.125
N220 G2 X–1.125 Y–.125 I–1.25 J0.0
N230 G1 X–3.8125
N240 G2 X–4.625 Y.6875 I0.0 J.813
N250 G80 G00 Z.1 M09
N260 G91 G28 Z0.0
N270 G28 X0.0 Y0.0
N280 M30
```

Programming Exercise 4-5 Answer

The tool path for the part contour is programmed using linear and circular interpolation and rapid positioning (G01, G02, G03 and G00), for the Machining Center. Radius programming is given using the program words R, I and J. Refer to Figure 43.

In this case, a .375 inch diameter end mill is used due to the three .1875 inch fillet radii required on the drawing.

Tool = .375 inch diameter, 2 flute, Carbide, End Mill
Cutting Speed = 600-2000
r/min = 6111-20372
in/tooth = .008-.015
in/min = 98.0-611.0

In this case, since the maximum r/min of the machine being used is 6000, this will be the r/min used in the program and the feed rates are adjusted accordingly.

The following drawing indicates the necessary tool radius offset for the correct tool path in relation to the angled surface and the counterclockwise arc.

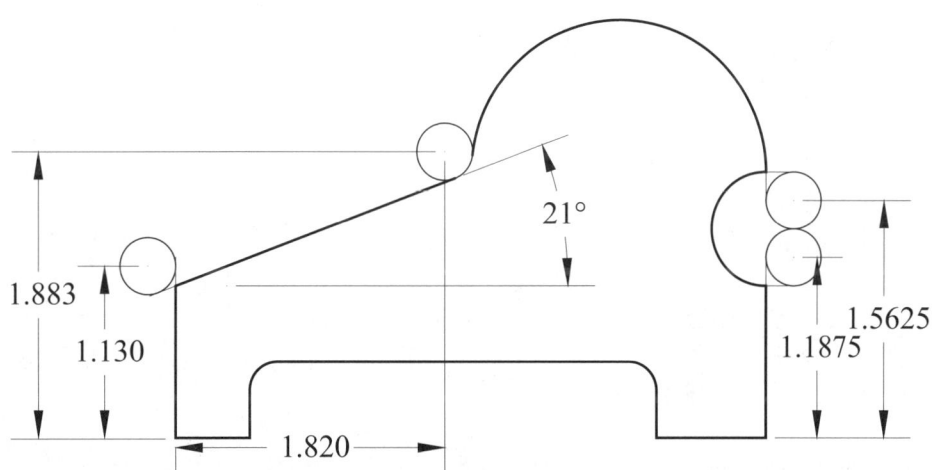

Figure A4-2   CNC Machining Center Programming Exercise 4-5, Cutter Location Coordinates

The following program is written entirely using the R command for the radii.

O00005
(CNC Machining Center Exercise #5)
(Date, By)
N10 G90 G20 G80 G40 G49
(Tool #1 .375 inch diameter, 2-flute Carbide, End Mill)
N20 T01 M06
N30 S6000 M03

N40 G54 G00 X–.1875 Y–.2875
N50 G43 Z1.0 H01
N60 Z.1 M08
N70 G1 Z–.25 F10.0
N80 Y1.13 F138.0
N90 X1.82 Y1.883
N100 G2 X4.1875 Y1.75 R1.1875
N110 G1 Y1.5625
N120 X4.0
N130 G3 Y1.1875 R.1875
N140 G1 X4.1875
N150 Y–.1875
N160 X3.0625
N170 Y.3125
N180 X.6875
N190 Y–.1875
N200 X–.2875
N210 G00 Z.1
N220 X–.1875 Y–.2875
N230 G1 Z–.5 F10.0
N240 Y1.13 F138.0
N250 X1.82 Y1.883
N260 G2 X4.1875 Y1.75 R1.1875
N270 G1 Y1.5625
N280 X4.0
N290 G3 Y1.1875 R.1875
N300 G1 X4.1875
N310 Y–.1875
N320 X3.0625
N330 Y.3125
N340 X.6875
N350 Y–.1875
N360 X–.2875
N370 G00 Z.1
N380 X–.1875 Y–.2875
N390 G80 Z.1 M09
N400 G91 G28 Z0.0
N410 G28 X0.0 Y0.0
N420 M30

The following program is written entirely using the I and J commands for the radii.

```
O0005
(CNC Machining Center Exercise #5)
(Date, By)
N10 G90 G20 G80 G40 G49
(Tool #1 .375 inch diameter, 2 flute, Carbide, End Mill)
N20 T01 M06
N30 S6000 M03
N40 G54 G00 X–.1875 Y–.2875
N50 G43 Z1.0 H01
N60 Z.1 M08
N70 G1 Z–.25 F10.0
N80 Y1.13 F138.0
N90 X1.82 Y1.883
N100 G2 X4.1875 Y1.75 I1.1875 J0.0
N110 G1 Y1.5625
N120 X4.0
N130 G3 Y1.1875 I0.0 J–.1875
N140 G1 X4.1875
N150 Y–.1875
N160 X3.0625
N170 Y.3125
N180 X.6875
N190 Y–.1875
N200 X–.2875
N210 G00 Z.1
N220 X–.1875 Y–.2875
N230 G1 Z–.5 F10.0
N240 Y1.13 F138.0
N250 X1.82 Y1.883
N260 G2 X4.1875 Y1.75 I1.1875 J0.0
N270 G1 Y1.5625
N280 X4.0
N290 G3 Y1.1875 I0.0 J–.1875
N300 G1 X4.1875
N310 Y–.1875
N320 X3.0625
N330 Y.3125
```

N340 X.6875
N350 Y–.1875
N360 X–.2875
N370 G00 Z.1
N380 X–.1875 Y–.2875
N390 G80 Z.1 M09
N400 G91 G28 Z0.0
N410 G28 X0.0 Y0.0
N420 M30

Circle Milling Answers

Programming Exercise 4-6 Answer

The tool path for the part contour is programmed using linear and circular interpolation and rapid positioning (G01, G02, G03 and G00), for the Machining Center. Radius programming is given using the program words R, I and J. Refer to Figure 4-7.

Figure A4-3 shows that a .75 inch diameter end mill is necessary to accomplish the axial cut in one step.

The following program is written entirely using the R command for the radii.

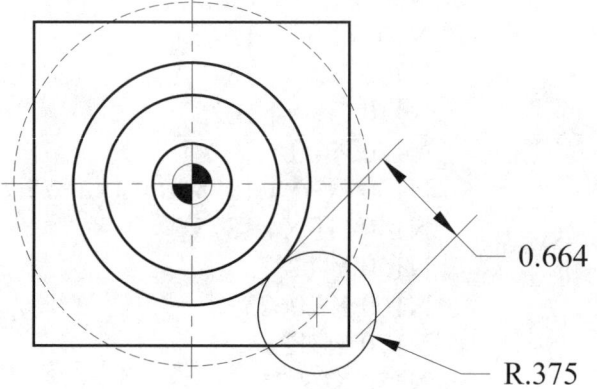

Figure A4-3   CNC Machining Center Programming Exercise 4-6, Cutter Size Requirement Calculation

Tool = .75 inch diameter, 4-flute HSS, End Mill
Cutting Speed = 25–140
r/min = 127–713
in/tooth = .001–.004
in/min = .508–11.4

O0006
(CNC Machining Center Exercise #6)
(Date, By)
N10 G90 G20 G80 G40 G49
(Tool #1 .75 inch diameter, 4-flute HSS, End Mill)
N20 T01 M06
N30 S420 M3
N40 G54 G00 X1.375 Y0.0
N50 G43 Z.1 H1

N60 Z.1 M08
N70 G01 Z–.375 F10.0
N80 X1.125
N90 G2 X–1.125 R1.125 F5.96
N100 X1.125 R1.125
N110 G00 Z.1
N120 X0.0 Y0.0
N130 G1 Z–.5 F10.0
N140 X.175
N150 G3 X–.175 R.175 F5.96
N160 X.175 R.175
N170 G80 G00 Z.1 M09
N180 G91 G28 Z0.0
N190 G28 X0.0 Y0.0
N200 M30

The following program is the same program written entirely using the I and J commands for the radii.

O0006
(CNC Machining Center Exercise #6)
(Date, By)
N10 G90 G20 G80 G40 G49
(Tool #1 .75 inch diameter, 4-flute HSS, End Mill)
N20 T01 M06
N30 S420 M3
N40 G54 G00 X1.375 Y0.0
N50 G43 Z.1 H1
N60 Z.1 M08
N70 G01 Z–.375 F10.0
N80 X1.125
N90 G2 X–1.125 I–1.125 J0.0 F5.96
N100 X1.125 I1.125 J0.0
N110 G0 Z.1
N120 X0.0 Y0.0
N130 G1 Z–.5 F10.0
N140 X.175
N150 G3 X–.175 I–.175 J0.0 F5.96
N160 X.175 I.175 J0.0
N170 G80 G0.0 Z.1 M09
N180 G91 G28 Z0.0
N190 G28 X0.0 Y0.0
N200 M30

Cutter Diameter Compensation Answers

The tool path for the part contour is programmed for Machining Center exercises 4-3, 4-4, 4-5, and 4-6 with (G41, G42, and G40) cutter diameter compensation. Radius programming is given using the program words R, I, and J.

Programming Exercise 4-7 Answer

Cutter Diameter Compensation for Exercise 4-3

Tool = .375 inch diameter, 4 flute, High Speed Steel, End Mill
Cutter Diameter Compensation offset # = D1
Cutting Speed = 5–85
r/min = 51–865
in/tooth = .001–.004
in/min = .20–14.0

```
O0007
(CNC Machining Center Exer-
   cise #7)
(Date, By)
N10 G90 G20 G80 G40 G49
(Tool #1 .375 inch diameter, 4-flute
   High Speed Steel, End Mill)
(Cutter Diameter Compensation
   offset #D1)
N20 T01 M06
N30 S458 M03
N40 G54 G00 X–.2 Y.2
N50 G43 Z1.0 H01
N60 Z.1 M08
N70 G1 Z–.125 F10.0
N80 X0.0 Y–.187 G41 D1
N90 X5.813 F7.0
N100 Y–2.844
N110 X.187
N120 Y.187
N130 G00 Z.1
N140 G0.0 X–.2 Y.2
N150 G1 Z–.25 F10.0
N160 X0.0 Y–.187 G41 D1
```

```
N170 X5.813 F7.0
N180 Y–2.844
N190 X.187
N200 Y.187
N210 G00 Z.1
N220 G80 Z.1 M09
N230 G40 Y.375
N240 G91 G28 Z0.0
N250 G28 X0.0 Y0.0
N260 M30
```

Programming Exercise 4-8 Answer

Cutter Diameter Compensation for Exercise 4-4

Tool = 1.00 inch diameter, 4 flute, High Speed Steel, End Mill
Cutter Diameter Compensation offset # = D31
Cutting Speed = 5–85
r/min = 19–324
in/tooth = .001–.004
in/min = .076–5.2

The following program is written entirely using the R command for the radii.

O0008
(CNC Machining Center Exercise #8)
(Date, By)
N10 G90 G20 G80 G40 G49
(Tool #1 1.0 inch diameter, 4-flute
   High Speed Steel, End Mill)
(Cutter Diameter Compensation
   offset #D1)
N20 T01 M06
N30 S172 M03
N40 FG00 G54 X–4.75 Y–.6
N50 G43 Z1.0 H01
N60 Z.1 M08
N70 G01 Z–.1875 F10.0
N80 X–4.125 Y0.0 G41 D1
N90 Y3.25 F2.6
N100 G2 X–3.875 Y3.5 R.25
N110 G1 X–.937
N120 G2 X–.375 Y2.938 R.562
N130 G1 Y1.125
N140 G2 X–1.125 Y.375 R.75
N150 G1 X–3.8125
N160 G2 X–4.125 Y.6875 R.3125
N170 G1 Z–.375
N180 Y3.25
N190 G2 X–3.875 Y3.5 R.25
N200 G1 X–.937
N210 G2 X–.375 Y2.938 R.562

N220 G1 Y1.125
N230 G2 X–1.125 Y.375 R.75
N240 G1 X–3.8125
N250 G2 X–4.125 Y.6875 R.3125
N260 G80 G0.0 Z.1 M09
N270 G40 X–4.75 Y0.0
N280 G91 G28 Z0.0
N290 G28 X0.0 Y0.0
N300 M30

The same program written entirely using the I and J commands for the radii.

```
O0008
(CNC Machining Center Exercise #8)
(Date, By)
N10 G90 G20 G80 G40 G49
(Tool #1 1.0 inch diameter, 4-flute, High Speed Steel, End Mill)
(Cutter Diameter Compensation offset #D1)
N20 T01 M06
N30 S172 M03
N40G00 G54 X–4.75 Y–.6
N50 G43 Z1.0 H01
N60 Z.1 M08
N70 G01 Z–.1875 F10.0
N75 X–4.125 Y0.0 G41 D1
N80 Y3.25 F2.6
N90 G2 X–3.875 Y3.5 I.25 J0.0
N100 G1 X–.937
N110 G2 X–.375 Y2.938 I0.0 J–.562
N120 G1 Y1.125
N130 G2 X–1.125 Y.375 I–.75J0
N140 G1 X–3.8125
N150 G2 X–4.125 Y.688 I0.0 J.3125
N160 G1 Z–.375
N170 Y3.25
N182 G2 X–3.875 Y3.5 I.25 J0.0
N190 G1 X–.937
N200 G2 X–.375 Y2.938 I0.0 J–.562
N210 G1 Y1.125
N220 G2 X–1.125 Y.375 I–.75 J0.0
N230 G1 X–3.8125
N240 G2 X–4.125 Y.688 I0.0 J.3125
N250 G80 G00 Z.1 M09
N260 G40 X–4.75 Y0.0
N260 G91 G28 Z0.0
N270 G28 X0.0 Y0.0
N280 M30
```

Programming Exercise 4-9 Answer

Cutter Diameter Compensation for Exercise 4-5

Tool = .375 inch diameter, 2 flute, Carbide, End Mill
Cutter Diameter Compensation offset # = D1
Cutting Speed = 600–2000
r/min = 6111–20,372
in/tooth = .008–.015
in/min = 98.0–611.0

The following program is written entirely using the R command for the radii.

```
O0009
(CNC Machining Center Exercise #9)
(Date, By)
N10 G90 G20 G80 G40 G49
(Tool #1 .375 inch diameter, 2-flute, Carbide, End Mill)
(Cutter Diameter Compensation offset #D1)
N20 T01 M06
N30 S6000 M03
N40 G00 G54 X–.2 Y–.2
N50 G43 Z1.0 H01
N60 Z.1 M08
N70 G1 Z–.25 F10.0
N80 X0.0 Y0.0 G41 D1
N90 Y1.0 F138.0
N100 X2.0 Y1.75
N110 G2 X4.0 Y1.75 R1.0
N120 G3 Y1.0 R.375
N130 G1 Y0.0
N140 X3.25
N150 Y.50
N160 X.50
N170 Y0.0
N180 X–.1875
N190 G0.0 Z.1
N200 X–.3 Y–.3
N210 G1 Z–.5 F10.0
N220 X0.0 Y0.0 G41 D1
N230 Y1.0 F138.0
N240 X2.0 Y1.75
```

N250 G2 X4.0 Y1.75 R1.0
N260 G3 Y1.0 R.375
N270 G1 Y0.0
N280 X3.25
N290 Y.5
N300 X.5
N310 Y0.0
N320 X0.0
N330 G00 Z.1
N340 G40 X–.4
N350 G80 Z.1 M09
N360 G91 G28 Z0.0
N370 G28 X0.0 Y0.0
N380 M30

The same program is written entirely using the I and J commands for the radii.

O0009
(CNC Machining Center Exer-
   cise #5)
(Date, By)
N10 G90 G20 G80 G40 G49
(Tool #1 .375 inch diameter,
   2-flute Carbide, End Mill)
(Cutter Diameter Compensation
   offset #D1)
N20 T01 M06
N30 S6000 M03
N40 G00 G54 X–.2 Y–.2
N50 G43 Z1.0 H01
N60 Z.1 M08
N70 G1 Z–.25 F10.0
N80 X0.0 Y0.0 G41 D1
N90 Y1.0 F138.0
N100 X2.0 Y1.75
N110 G2 X4.0 Y1.75 I1.0 J0.0
N120 G3 Y1.0 I0.0 J–.375
N130 G1 Y0.0
N140 X3.25
N150 Y.50

N160 X.50
N170 Y0.0
N180 X0.0
N190 G00 Z.1
N200 X–.4 Y–.4
N220 Z–.5 F10.0
N210 G1 X0.0 Y0.0
N230 Y1.0 F138.0
N240 X2.0 Y1.75
N250 G2 X4.0 Y1.75 I1.0 J0.0
N260 G3 Y1.0 I0.0 J–.375
N270 G1 Y0.0
N280 X3.25
N290 Y.5
N300 X.5
N310 Y0.0
N320 X0.0
N330G0Z.1
N340 G40 X–.4
N330 G80 Z.1 M09
N350 G91 G28 Z0.0
N360 G28 X0.0 Y0.0
N370 M30

Programming Exercise 4-10 Answer

Cutter Diameter Compensation for Exercise 4-6

Tool = .75 inch diameter, 4-flute HSS, End Mill
Cutting Speed = 25–140
r/min = 127–713
in/tooth = .001–.004
in/min = .508–11.4

The following program is written entirely using the R command for the radii.

```
O0010
(CNC Machining Center Exercise #10)
(Date, By)
N10 G90 G20 G80 G40 G49
(Tool #1 .75 inch diameter, 4-flute HSS, End Mill)
(Cutter Diameter Compensation offset #D1)
N20 T01 M06
N30 S420M3
N40 G54G0X1.5Y0
N50 G43Z.1H1
N60 Z.1 M08
N70 G01 Z–.375 F10.0
N80 X.75 G41 D1 F5.96
N90 G2 X–.75 R.75
N100 X.75 R.75
N110 G00 G40 Z.1
N120 X0.0 Y0.0
N130 G1 Z–.5 F10.0
N140 X.55 Y0.0 G41 D1
N150 G3 X–.55 R.55
N160 X.55 R.55
N170 G80 G00 Z.1 M09
N180 G40 X–.1
N190 G91 G28 Z0.0
N200 G28 X0.0 Y0.0
N210 M30
```

The same program is written entirely using the I and J commands for the radii.

O0010
(CNC Machining Center Exer-
   cise #10)
(Date, By)
N10 G90 G20 G80 G40 G49
(Tool #1 .75 inch diameter,
   4-flute HSS, End Mill)
(Cutter Diameter Compensation
   offset #D1)
N20 T01 M06
N30 S420 M3
N40 G00 G54 X1.5 Y0.0
N50 G43Z.1 H1
N60 Z.1 M08
N70 G01 Z–.375 F10.0
N80 X.75 G41 D1 F5.96
N90 G2 X–.75 I–.75 J0.0
N100 X.75 I.75 J0.0
N110 G40 G00 Z.1
N120 X0.0 Y0.0

N130 G1 Z–.5F10.0
N140 X.55 G41 D1 F5.96
N150 G3 X–.55 I–.55 J0.0
N160 X.55 I.55 J0.0
N170 G80 G00 Z.1 M09
N180 G40 X–.1
N190 G91 G28 Z0.0
N200 G28 X0.0 Y0.0
N210 M30

Canned Cycles Answers

Programming Exercise 4-11 Answer

G81 Drilling

The tool path for the drilled hole is programmed using Canned Cycle (G81 and G00) for the Machining Center. This program segment could be added to the beginning of the program for CNC Machining Center Exercise 4-6 to aid the entry of the end mill.

Calculation is necessary to allow for the drill point.

.25 × TAN 31° = .150

.500 – .150 = .350 for the drill depth

Tool = .5 inch diameter, HSS, Drill
Cutting Speed = 25–140
r/min = 190–1070
in/tooth = .001–.004
in/min = .38–8.5

O0011
(CNC Machining Center Exercise #11)
(Date, By)
N10 G90 G20 G80 G40 G49
(Tool #1 .#5 Center Drill)
N20 T01 M06
N30 S630 M03
N40 G00G54 X0.0 Y0.0
N50 G43 Z1.0 H01
N60 Z.1 M08
N70 G81 G99 Z–.25 R.1 F4.0
N80 G80 Z.1 M09
N90 G91 G28 Z0.0
N100 G28 X0.0 Y0.0
(Tool #2 .5 inch diameter, HSS, Drill)
N110 T01 M06
N120 S630 M03
N130 G00 G54 X0.0 Y0.0
N140 G43 Z1.0 H02
N150 Z.1 M08
N160 G81 G99 Z–.35 R.1 F4.0
N170 G80 Z.1 M09
N180 G91 G28 Z0.0
N190 G28 X0.0 Y0.0
(Tool #3 .75 inch diameter, 4-flute HSS, End Mill)
N200 T03 M06
N210 S420 M3
N220 G00 G54 X1.375 Y0.0
N230 G43 Z.1 H3
N240 Z.1 M08
N250 G01 Z–.375 F10.0
N260 X1.125
N270 G2 X–1.125 R1.125 F5.96
N280 X1.125 R1.125
N290 G00 Z.1
N300 X0.0 Y0.0
N310 G1 Z–.5 F10.0

N320 X.175
N330 G3 X–.175 R.175 F5.96
N340 X.175 R.175
N350 G00 G80 Z.1 M09
N360 G91 G28 Z0.0
N370 G28 X0.0 Y0.0
N380 M30

Programming Exercise 4-12 Answer

G81 and G73 Drilling

The holes are added to the existing program for CNC Machining Center Exercise 4-7, using Canned Drilling Cycles (G81, G73, and G80). Refer to Figure 4-9.

Tool 1 = .375 diameter, 4-flute HSS End Mill
Cutter Diameter Compensation offset #D1
Cutting Speed = 5–85
r/min = 51–865
in/tooth = .001–.004
in/min = .20–14.0

Tool 2 = .75 diameter 90° HSS Spot Drill
Cutting Speed = 5–85
r/min = 25–433
in/tooth = .001–.004
in/min = .0025–1.73

Tool 3 = .25 diameter HSS Drill
Cutting Speed = 5–85
r/min = 76–1299
in/tooth = .001–.004
in/min = .152–10.4

Tool 4 = .375 diameter HSS Drill
Cutting Speed = 5–85
r/min = 51–865
in/tooth = .001–.004
in/min = .20–14.0

Calculation required for the Spot Drill depth on the .25 diameter hole.
.280/2 = .140 × TAN 45° = .140

Calculation required for the Spot Drill depth on the .375 diameter hole.
.405/2 = .2025 × TAN 45° = .2025

Calculation required for the drill point on the .25 diameter hole.
.250/2 = .125 × TAN 31° = .075

Calculation required for the drill point on the .375 diameter hole.
.375/2 = .1875 × TAN 31° = .113

O0012
(CNC Machining Center Exercise #12)
(Date, By)
N10 G90 G20 G80 G40 G49
(Tool #1 .375 inch diameter, 4-flute High Speed Steel, End Mill)
(Cutter Diameter Compensation offset #D1)
N20 T01 M06
N30 S458 M03
N40 G00 G54 X–.2 Y.2
N50 G43 Z1.0 H01
N60 Z.1 M08
N70 G1 Z–.125 F10.0
N80 X0.0 Y–.187 G41 D1
N90 X5.813 F7.0
N100 Y–2.844
N110 X.187
N120 Y.187
N130 G00 Z.1
N140 X–.2 Y.2
N150 G1 Z–.25 F10.0
N160 X0.0 Y–.187 G41 D1
N170 X5.813 F7.0
N180 Y–2.844
N190 X.187
N200 Y.187
N210 G00 Z.1
N220 G00 G80 Z.1 M09
N230 G40 Y.375
N240 G91 G28 Z0.0
N250 G28 X0.0 Y0.0
N260 M01
(Tool 2 = .75 diameter 90° HSS Spot Drill)
N270 T02 M06
N280 S230 M03
N290 G90 G20 G80 G40 G49

N300 G00 G54 X.75 Y–.75
N310 G43 Z1.0 H02
N320 Z.1 M08
N330 G81 G99 Z–.14 R.1 F1.0
N340 X5.281 Y–2.281
N350 G81 G99 X5.531 Y–.781 Z-.2025 R.1
N360 X.5 Y–3.25
N370 G80 Z.1 M09
N380 G91 G28 Z0.0
N390 M01
(Tool 3 = .25 diameter HSS Drill)
N400 T03 M06
N410 S688 M03
N420 G90 G20 G80 G40 G49
N430 G54 G0.0 X.75 Y–.75
N440 G43 Z1.0 H03
N450 Z.1 M08
N460 G73 G99 Z–.825 Q.17 R.1 F3.44
N470 X5.281 Y–2.281
N480 G80 Z.1 M09
N490 G91 G28 Z0.0
N500 M01
(Tool 4 = .375 diameter HSS Drill)
N510 T04 M06
N520 S458 M03
N530 G90 G20 G80 G40 G49
N540 G00 G54 X5.531 Y–.781
N550 G43 Z1.0 H03
N560 Z.1 M08
N570 G73 G99 Z–.863 Q.25 R.1 F7.0
N580 X.5 Y–.25
N590 G80 Z.1 M09
N600 G91 G28 Z0.0
N610 G28 X0.0 Y0.0
N620 M30

Programming Exercise 4-13 Answer

G81, G83, and G82 Drilling

The holes are added to the existing program for CNC Machining Center Exercise 4-8, using Canned Drilling Cycles (G81, G83, G82, and G80). Refer to Figure 4-10.

> Tool 1 = 1.00 inch diameter, 4-flute High Speed Steel, End Mill
> Cutter Diameter Compensation offset # = D1
> Cutting Speed = 5–85
> r/min = 19–324
> in/tooth = .001–.004
> in/min = .076–5.2

> Tool 2 = #6 Center Drill
> Cutting Speed = 5–85
> r/min = 38–650
> in/tooth = .001–.004
> in/min = .076–5.2

> Tool 3 =.375 inch diameter, HSS, Drill
> Cutting Speed = 5–85
> r/min = 51–866
> in/tooth = .001–.004
> in/min = .102–7.0

A calculation is necessary to allow addition for the drill point.

$.1875 \times TAN\ 31° = .113$

$1.0 + .113 = 1.13$ for the drill depth

> Tool 4 =.50 inch diameter, 4-flute HSS, End Mill
> Cutting Speed = 5–85
> r/min = 19–324
> in/tooth = .001–.004
> in/min = .076–5.2

O0013
(CNC Machining Center Exercise #13)
(Date, By)
N10 G90 G20 G80 G40 G49
(Tool #1 1.0 inch diameter, 4-flute High Speed Steel, End Mill)
(Cutter Diameter Compensation offset #D1)
N20 T01 M06
N30 S172 M03
N40 G00 G54 X–4.75 Y–.6
N50 G43 Z1.0 H01
N60 Z.1 M08
N70 G01 Z–.1875 F10.0
N80 X–4.125 Y0.0 G41 D1
N90 Y3.25 F2.6
N100 G2 X–3.875 Y3.5 R.25
N110 G1 X–.937
N120 G2 X–.375 Y2.938 R.562
N130 G1 Y1.125
N140 G2 X–1.125 Y.375 R.75
N150 G1 X–3.8125
N160 G2 X–4.125 Y.6875 R.3125
N170 G1 Z–.375
N180 Y3.25
N190 G2 X–3.875 Y3.5 R.25
N200 G1 X–.937
N210 G2 X–.375 Y2.938 R.562
N220 G1 Y1.125
N230 G2 X–1.125 Y.375 R.75
N240 G1 X–3.8125
N250 G2 X–4.125 Y.6875 R.3125
N260 G00G80 Z.1 M09
N270 G40 X–4.75 Y0.0
N280 G91 G28 Z0.0
N290 G28 X0.0 Y0.0
N290 M01
(Tool 2 #6 Center Drill)
N300 T01 M06
N310 S344 M03

N320 G90 G20 G80 G40 G49
N330 G00 G54 X–1.125 Y1.125
N340 G43 Z1.0 H02
N350 Z.1 M08
N360 G81 G99 Z–.4 R.1 F1.72
N370 X–.937 Y2.938
N380 X–3.875
N390 X–3.812 Y.688
N400 G80 Z.1 M09
N410 G91 G28 Z0.0
N420 M01
(Tool 3 .375 inch diameter, HSS, Drill)
N430 T03 M06
N440 S459M03
N450 G90 G20 G80 G40 G49
N460 G00G54 X–1.125 Y1.125
N470 G43 Z1.0 H03
N480 Z.1 M08
N490 G83 G99 Z–1.13 Q.25 R.1 F2.3
N500 X–.937 Y2.938
N510 X–3.875
N520 X–3.812 Y.688
N530 G80 Z.1 M09
N540 G91 G28 Z0.0
N550 M01
(Tool 4 .50 inch diameter, 4-flute HSS End Mill)
N560 T04 M06
N570 S172 M03
N580 G90 G20 G80 G40 G49
N590 G00 G54 X–1.125 Y1.125
N600 G43 Z1.0 H03
N610 Z.1 M08
N620 G82 G99 Z–.375 P200 R.1 F2.6
N630 X–.937 Y2.938
N640 X–3.875
N650 X–3.812 Y.688
N660 G80 Z.1 M09
N670 G91 G28 Z0.0
N680 G28 X0.0 Y0.0
N690 M30

Programming Exercise 4-14 Answer

G81, G82, G83, and G84

The holes are programmed using Canned Drilling Cycles (G81, G82, G83, G84, and G80). Refer to Figure 4-11.

> Tool 1 = #5 Center Drill
> Cutting Speed = 165–850
> r/min = 1440–7421
> in/tooth = .002–.006
> in/min = 5.8–89.0

Calculations are necessary to allow addition for the Center Drill point to allow for a .395 inch diameter countersink for the thread lead.

> $.0938 \times TAN\ 30° = .054$ (for the 120° tip)
> $.1038 \times TAN\ 60° = .1798$ (for the 60° portion)
> $.054 + .1798 + .1875 = .4213$

> Tool 2 = 5/16 inch diameter, HSS, Drill
> Cutting Speed = 165–850
> r/min = 2017–10,390
> in/tooth = .002–.006
> in/min = .8.0–125.0

A calculation is necessary to allow addition for the drill point.

> $.1563 \times TAN\ 31° = .0939$

> $1.0 + .0939 = 1.0939$ for the drill depth

> Tool 3 = 3/8-16 TAP
> Cutting Speed = 85
> r/min = 865

A calculation for the feed is necessary and considers the use of a floating-type tap holder.

> $865 \times 1/16 = 54.0$

PROCEED WITH CAUTION

Tool 4 =.375 inch diameter, HSS, Drill
Cutting Speed = 165–850
r/min = 1681–8658
in/tooth = .002–.006
in/min = 6.7–104.0

A calculation is necessary to allow addition for the drill point.

.1875 × TAN 31° = .113

1.0 + .113 = 1.13 for the drill depth

Tool 5 =.5625 inch diameter, HSS, End Mill
Cutting Speed = 165–850
r/min = 1120–5772
in/tooth = .002–.006
in/min = 4.5–69.0

In Figure A4-4, the coordinates for the Bolt Hole Circle are given.

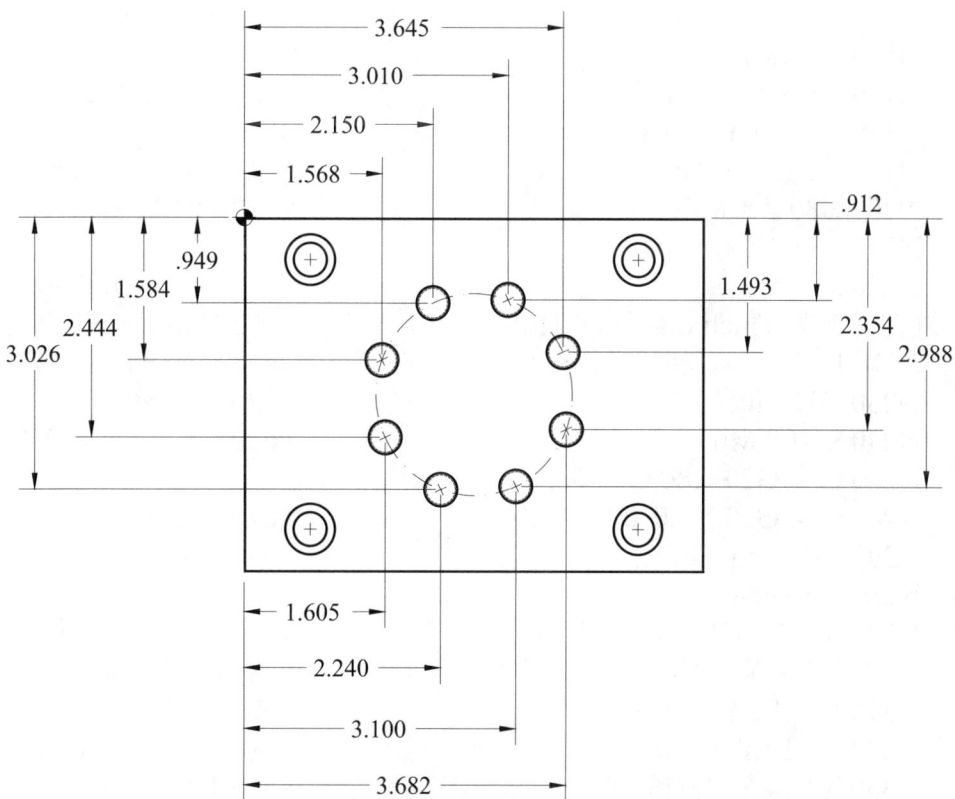

Figure A4-4   CNC Machining Center Programming Exercise 4-14 Coordinates

O0014
(CNC Machining Center Exercise #14)
(Date, By)
N10 G90 G20 G80 G40 G49
(Tool 1 #5 Center Drill)
N20 T01 M06
N30 S4431 M03
N40 G90 G20 G80 G40 G49
N50 G00 G54 X.75 Y–.469
N60 G43 Z1.0 H01
N70 Z.1 M08
N80 G81 G99 Z–.4213 R.1 F47.0
N90 Y–3.469
N100 X4.5
N110 Y–.469
N120 X3.010 Y–.912
N130 X3.645 Y–1.493
N140 X3.682 Y–2.354
N150 X3.10 Y–2.988
N160 X2.24 Y–3.026
N170 X1.605 Y–2.444
N180 X1.568 Y–1.584
N190 X2.15 Y–.949
N200 G80 Z.1 M09
N210 G91 G28 Z0.0
N220 M01
(Tool 2 5/16 inch diameter, HSS, Drill)
N230 T02 M06
N240 S6000 M03
N250 G90 G20 G80 G40 G49
N260 G54 G00 X3.010 Y–.912
N270 G43 Z1.0 H02
N280 Z.1 M08
N290 G83 G99 Z–1.094 R.1 F47.0
N300 X3.645 Y–1.493
N310 X3.682 Y–2.354
N320 X3.10 Y–2.988
N330 X2.24 Y–3.026
N340 X1.605 Y–2.444

N350 X1.568 Y–1.584
N360 X2.15 Y–.949
N370 G80 Z.1 M09
N380 G91 G28 Z0.0
N390 M01
(Tool 3 3/8-16 TAP)
N400 T03 M06
N410 S865 M03
N420 G90 G20 G80 G40 G49
N430 G00 G54 X3.010 Y–.912
N440 G43 Z1.0 H03
N450 Z.1 M08
N460 G84 G99 Z–1.2 R.1 F54.0
N470 X3.645 Y–1.493
N480 X3.682 Y–2.354
N490 X3.10 Y–2.988
N500 X2.24 Y–3.026
N510 X1.605 Y–2.444
N520 X1.568 Y–1.584
N530 X2.15 Y–.949
N540 G80 Z.1 M09
N550 G91 G28 Z0.0
N560 M01
(Tool 4 3/8 diameter drill)
N570 T04 M06
N580 S5169 M03
N590 G90 G20 G80 G40 G49
N600 G00 G54 X.75 Y–.469
N610 G43 Z1.0 H04
N620 Z.1 M08
N630 G83 G99 Z–1.13 Q.25 R.1 F41.0
N640 Y–3.469
N650 X4.5
N660 Y–.469
N670 G80 Z.1 M09
N680 G91 G28 Z0.0
N690 M01
(Tool 5 9/16 2-flute End Mill)
N700 T05 M06
N710 S3446 M03

N720 G90 G20 G80 G40 G49
N730 G00 G54 X.75 Y–.469
N740 G43 Z1.0 H05
N750 Z.1 M08
N760 G82 G99 Z–.25 P200 R.1
　　F28.0
N770 Y–3.469
N780 X4.5

N790 Y–.469
N800 G80 Z.1 M09
N810 G91 G28 Z0.0
N820 G80 Z.1 M09
N830 G91 G28 Z0.0
N840 G28 X0.0 Y0.0
N850 M30

CNC Machining Center Combined Projects Answers

Programming Exercise 4-15 Answer

The following is the program code for the Machining Center part in the Process Planning section of this workbook. The face mill tool path is programmed without the use of cutter radius compensation. Refer to Figure 4-12.

Because the finished surface must be zero, we must touch off the raw material and then set the Z-offset value at –.100 inch.

Tool 1 = Face Mill, 3.0 inch diameter, 90°, 5 teeth, Carbide)
Cutting Speed = 755–1720
r/min = 961–2190
in/tooth = .020–.039
in/min = 96.0–335.0

Tool 2 = 1.0 inch diameter, HSS, 2-Flute End Mill
Cutting Speed = 165–850
r/min = 630–3247
in/tooth = .002–.006
in/min = 2.5–39.0

Tool 3 = #5 HSS Center Drill
Cutting Speed = 165–850
r/min = 1261–6494
in/tooth = .002–.006
in/min = 5.0–78.0

Tool 4 = .4219 inch (27/64) diameter HSS Drill
Cutting Speed = 165–850
r/min = 1494–7696
in/tooth = .002–.006
in/min = 6.0–92.0

Calculation required for the drill point on the .4219 inch diameter hole.

.211 × TAN 31° = .1268

1.125 + .1268 = 1.2518

Tool 5 = .4375 diameter HSS Reamer
Cutting Speed = 203
r/min = 1772
in/tooth = .003
in/min = 32.0

```
O0015
(CNC Machining Center Exercise #14)
(Date, By)
N10 G90 G20 G80 G40 G49
(Tool #1 Face Mill 3.0 inch diameter 90° 5 teeth Carbide)
N20 T01 M6
N30 G00 G54 X3.725 Y1.0
N40 S1238 M3
N50 G43 Z1.0 H01
N60 Z.2 M08
N70 G01 Z0.0 F10.0
N80 X–3.725 F216.0
N90 Z.2
N100 X3.725 Y–1.0
N110 Z0.0
N120 X–3.725
N130 Z.1
N140 X3.5167 Y0.0
N150 Z–.1875 F10.0
N160 X1.7583 Y–3.0455 F216.0
N170 X–1.7583
N180 X–3.5167 Y0.0
N190 X–1.7583 Y3.0455
N200 X1.7583
N210 X3.5167 Y0.0
N220 Z–.375 F10.0
N230 X1.7583 Y–3.0455 F216.0
```

N240 X–1.7583
N250 X–3.5167 Y0.0
N260 X–1.7583 Y3.0455
N270 X1.7583
N280 X3.5167 Y0.0
N290 G80Z.1 M09
N300 G91 G28 Z0.0
N310 M01
(Tool #2 1.0 inch diameter, HSS, 2-Flute End Mill)
(Cutter Diameter Compensation offset #D2)
N320 T02 M6
N330 S1939 M3
N340 G90 G20 G80 G40 G49
N350 G00 G54 X2.7 Y.6
N360 G43 Z1.0 H02
N370 Z.1 M08
N380 G1 Z–.375 F10.0
N390 X1.75Y0 G41 D2
N400 X.875 Y–1.5155 F21.0
N410 X–.875
N420 X–1.75 Y0.0
N430 X–.875 Y1.5155
N440 X.875
N450 X1.75 Y0.0
N460 G00 G80 Z.1 M09
N470 G91 G28 Z0.0
N480 M01
(Tool #3 = #5 HSS Center Drill)
N490 T03 M6
N500 S3878 M03
N510 G90 G20 G80 G40 G49
N520 G00 G54 X.7071 Y.7071
N530 G43 Z.1 H03
N540 G81 Z–.25 R.1 F42.0

N550 Y–.7071
N560 X–.7071
N570 Y.7071
N580 G80 Z.1 M09
N590 G91 G28 Z0.0
N600 M01
(Tool 4 = .4219 27/64 diameter HSS Drill)
N610 T04 M6
N620 S4595 M03
N630 G90 G20 G80 G40 G49
N640 G00 G54 X.7071 Y.7071
N650 G43 Z.1 H04
N660 G83 Z–1.252 Q.282 R.1 F49.0
N670 Y–.7071
N680 X–.7071
N690 Y.7071
N700 G80 Z.1 M09
N710 G91 G28 Z0.0
N720 M01
(Tool #5 = .4375 diameter HSS Reamer)
N730 T05 M6
N740 S1772 M03
N750 G90 G20 G80 G40 G49
N760 G00 G54 X.7071 Y.7071
N770 G43 Z.1 H05
N780 G85 Z–1.252 R.1 F32.0
N790 Y–.7071
N800 X–.7071
N810 Y.7071
N820 G80 Z.1 M09
N830 G91 G28 Z0.0
N840 G28 X0.0 Y0.0
N850 M30

CNC Machining Center Subprogram Application Answers

Programming Exercise 4-16 Answer

M98 and M99 and G54–G59

The following is the program to machine all of the holes and pocket for Figure 4-13. Subprogram application (M98 and M99), multiple work offsets (G54–G59), and G68 for coordinate system rotation are used for the slots.

Tool 1 = .75 diameter 90° HSS Spot Drill
Cutting Speed = 165–850
r/min = 840–4329
in/tooth = .002–.006
in/min = 1.68–26.0

Tool 2 = 31/64 .4843 inch diameter HSS Drill
Cutting Speed = 165–850
r/min = 1301–6704
in/tooth = .002–.006
in/min = 5.2–80.4

Tool 3 = .375 inch diameter HSS Drill
Cutting Speed = 165–850
r/min = 1681–8658
in/tooth = .002–.006
in/min = 6.72–104.

Tool 4 = .3750 inch diameter HSS Reamer
Cutting Speed = 55
r/min = 560
in/tooth = .002
in/min = 6.7

Tool 5 = 5/8 inch diameter, HSS, 2-Flute End Mill
Cutting Speed = 165–850
r/min = 1008–5195
in/tooth = .002–.006
in/min = 4.0–62.3

Tool 6 = 1/8 inch diameter, HSS, 2-Flute End Mill
Cutting Speed = 165–850
r/min = 6000
in/tooth = .002
in/min = 24.0

Calculation required for the Spot Drill depth on the .5 diameter holes allowing .010 inch for burr removal.

.520/2 = .260 × TAN 45° = .260

Calculation required for the drill point on the 31/64 .4843 diameter hole.

.4843/2 = .2422 × TAN 31° = .1455

Calculation required for the drill point on the .375 diameter hole.

.375/2 = .1875 × TAN 31° = .113

O0016
(CNC Machining Center Exercise #16)
(Date, By)
N10 G90 G20 G80 G40 G49
(Tool 1 = .75 diameter 90° HSS Spot Drill)
N20 T01 M6
N30 G00 G54 X1.0 Y1.0
N40 S2585 M3
N50 G43 Z1.0 H01
N60 Z.1 M08
N70 G99 G81 Z–.26 R.1 F10.3
N80 X–1.0 Y–1.0
N90 Y1.0
N100 X1.0 Y–1.0
N110 G80 Z.1 M09
N120 G91 G28 Z0.0
N130 M01
(Tool 2 = 31/64 .4843 inch diameter HSS Drill)
N140 T02 M6
N150 G90 G20 G80 G40 G49

N160 G00 G54 X–1.0 Y1.0
N170 S4003 M3
N180 G43 Z1.0 H02
N190 Z.1 M08
N200 G99 G83 Z–.896 R.1 Q.323 F32.0
N210 X1.0 Y–1.0
N220 G80 Z.1 M09
N230 G91 G28 Z0.0
N240 M01
(Tool 3 = .375 inch diameter HSS Drill)
N250 T03 M6
N260 G90 G20 G80 G40 G49
N270 G00 G54 X1.0 Y1.0
N280 S3489 M3
N290 G43 Z1.0 H03
N300 Z.1 M08
N310 G99 G83 Z–.863 R.1 Q.25 F28.0
N320 X–1.0 Y–1.0
N330 X0.0 Y0.0
N340 G80 Z.1 M09

N350 G91 G28 Z0.0
N360 M01
(Tool 4 = .5000 inch diameter
  HSS Reamer)
N370 T03 M6
N380 G90 G20 G80 G40 G49
N390 G00 G54 X–1.0 Y1.0
N400 S560 M3
N410 G43 Z1.0 H04
N420 Z.1 M08
N430 G99 G86 Z–.875 R.1 F6.7
N440 X1.0 Y–1.0
N450 G80 Z.1 M09
N460 G91 G28 Z0.0
N470 M01
(Tool 5 = 5/8 inch diameter,
  HSS, 2 Flute End Mill)
N480 T05 M6
N490 G90 G20 G80 G40 G49
N500 G00 G54 X1.0 Y1.0
N510 S3102 M3N520 G43 Z1.0
  H05
N530 Z.1 M08
N540 G99 G82 Z–.5 R.1 P2.
  F24.8
N550 X–1.0 Y–1.0
N560 G80 Z.1
N570 X.0082 Y–.3334
N580 G1 Z–.25
N600 G3X.3335 Y0.0 R.3335
N610 G1 X–.3335
N620 G2 X–.0082 Y.3334
  R.3335
N630 G1 X.0082
N640 G0 Z.1
N650 X.3435 Y0.0
N660 G1 Z–.25
N670 G3 X0.0 Y.3435 R.3435
N680 X–.3435 Y0.0 R.3435
N690 X0.0 Y–.3435 R.3435
N700 X.3435 Y0.0 R.3435

N710 G00 Z.1
N720 G91 G28 Z0.0
N730 M01
(Tool 6 = 1/8 inch diameter,
  HSS, 2-Flute End Mill)
N740 T06 M6
N750 G90 G20 G80 G40 G49
N760 G00 G54 X.931 Y0.0
N770 G68 X0.0 Y0.0 R0.0
N780 S6000 M3
N790 G43 Z1.0 H06
N800 Z.1 M08
N810 G1 Z–.25 F24.0
N820 M98 P3456
N830 G00 Z.1
N840 G68 X0.0 Y0.0 R60.0
N850 G1 Z–.25 F6.16
N860 M98 P3456
N870 G00 Z.1
N880 G68 X0.0 Y0.0 R120.0
N890 M98 P3456
N900 G00 Z.1
N910 G68 X0.0 Y0.0 R180.0
N920 M98 P3456
N930 G00 Z.1
N940 G68 X0.0 Y0.0 R240.0
N950 M98 P3456
N960 G00 Z.1
N970 G68 X0.0 Y0.0 R300
N980 M98 P3456
N990 G69
N1000 G00 Z.1 M09
N1010 G91 G28 Z0.0
N1020 G28 X0.0 Y0.0 A0.0
N1030 M30

Subprogram for program
  #O0016
O3456
N1020 G3 X.8601 Y.3563 R.931
N1030 X.8315 Y.3754 R.031

N1040 X.8005 Y.3444 R.031
N1050 X.8029 Y.3326 R.031
N1060 G2 X.869 Y0.0 R.869
N1070 G3 X.9 Y–.031 R.031
N1080 X.931 Y0.0 R.031
N1090 M99

Programming Exercise 4-17 Answer

The following program is to machine all of the holes and step cutouts in the block in Figure 4-14, using the G98 command with Canned Cycles.

A 1.0 inch diameter 4-flute HSS End Mill could be used to rough out the steps, leaving .03 inch for finish. In this answer program, the 3.0 inch Face Mill is chosen instead. This program is also a good candidate for the use of a subprogram for repetition of the hole locations.

Tool 1 = Face Mill, 3.0 inch diameter, 90°, 5 teeth, Carbide)
Cutting Speed = 90–685
r/min = 115–872
in/tooth = .020–.039
in/min = 12.0–170.0

Tool 2 = 1.0 inch diameter, HSS, 4-Flute Roughing End Mill
Cutting Speed = 25–140
r/min = 95–535
in/tooth = .001–.004
in/min = .380–8.6

Tool 3 = 1.0 inch diameter, HSS, 4-Flute End Mill
Cutting Speed = 25–140
r/min = 95–535
in/tooth = .001–.00
4in/min = .380–8.6

Tool 4 = #5 HSS Center Drill
Cutting Speed = 25–140
r/min = 218–1222
in/tooth = .001–.004
in/min = .436–9.8

Tool 5 = #7 .201 inch diameter HSS Drill
Cutting Speed = 25–140
r/min = 475–2660
in/tooth = .001–.004
in/min = .950–21.0
Add a minimum of .200 inch depth to allow for the tap lead.

Tool 6 = 3/8 inch diameter HSS Drill
Cutting Speed = 25–140
r/min = 255–1426
in/tooth = .001–.004
in/min = .510–11.4

Tool 7 = 23/64 (.3594) inch diameter HSS Drill
Cutting Speed = 25–140
r/min = 266–1488
in/tooth = .001–.004
in/min = .532–11.9

Tool 8 = .3750 diameter HSS Reamer
Cutting Speed = 38
r/min = 387
in/tooth = .0025
in/min = 5.8

Tool 9 = 5/8 (.625) inch diameter, HSS, 4-Flute End Mill
Cutting Speed = 25–140
r/min = 153–856
in/tooth = .001–.004
in/min = .612–14.0

Tool 10 = 1/4-20 Tap
Cutting Speed = 25
r/min = 382
in/tooth = .002
in/min = 19.0
382 * .05 = 19.0

O0017
(CNC Machining Center Exercise #17)
(Date, By)
N10 G90 G20 G80 G40 G49
(Tool #1 Face Mill 3.0 inch diameter 90° 5 teeth Carbide)
N20 T01 M6
N30 G00 G54 X2.125 Y1.60
N40 S494 M3
N50 G43 Z1.0 H01
N60 Z.1 M08
N70 G01 Z–.25 F10.0
N80 Y–4.6 F18.0
N90 G00 Z.1
N100 Y1.6
N110 G01 Z–.50 F10.0
N120 Y–4.6 F18.0
N130 G80 Z.1 M09
N140 G91 G28 Z0.0
N150 M01
(Tool 2 = 1.0 inch diameter, HSS, 4 Flute Roughing End Mill)
N160 T02 M6
N170 G90 G20 G80 G40 G49
N180 G54 G00 X2.125 Y.60
N190 S315 M3
N200 G43 Z1.0 H02
N210 Z.1 M08
N220 G01 Z–.75 F20.0
N230 Y–3.6 F3.0
N240 G00 Z.1
N250 Y.6
N260 G01 Z–1.5 F20.0
N270 Y–3.6 F3.0
N280 G80 Z.1 M09
N290 G91 G28 Z0.0
N300 M01
(Tool 3 = 1.0 inch diameter, HSS, 4-Flute End Mill)

(Cutter Diameter Compensation #D3)
N310 T03 M6
N320 G90 G20 G80 G40 G49
N330 G00 G54 X1.0 Y.7
N340 S315 M3
N350 G43 Z1.0 H03
N360 Z.1 M08
N370 G01 Z–.5 F10.0
N380 X.5 Y.5 G41 D3
N390 Y–4.0 F3.0
N400 X3.75 F20.0
N410 Y.5 F3.0
N420 G1 Z–1.5 F10.0
N430 X1.5 G41 D3 F20.0
N440 Y–4.0 F3.0
N450 X2.75 F20.0
N460 Y.5 F3.0
N470 G80 G40 Z.1 M09
N480 G91 G28 Z0.0
N490 M01
(Tool 4 = #5 HSS Center Drill)
N500 T04 M6
N510 G90 G20 G80 G40 G49
N520 G00 G54 X.25 Y–.5
N530 S720 M3
N540 G43 Z1.0 H04
N550 Z.1 M08
N560 G81 G99 Z–.25 R.1 F3.6
N570 Y–1.0
N580 Y–1.5
N590 Y–2.0
N600 Y–2.5
N610 X4.0
N620 Y–2.0
N630 Y–1.5
N640 Y–1.0
N650 Y–.5
N660 G81 G98 X3.25 Y–1.0 Z–.75 R–.4
N670 Y–2.0

N680 X1.0
N690 Y–1.0
N700 G81 G98 X2.125 Y–.75
  Z–1.75 R–1.4
N710 Y–2.25
N720 G80 Z.1 M09
N730 G91 G28 Z0.0
N740 M01
(Tool 5 = #7 .201 inch diameter
  HSS Drill)
N750 T05 M6
N760 G90 G20 G80 G40 G49
N770 G00 G54 X.25 Y–.5
N780 S1566 M3
N790 G43 Z1.0 H05
N800 Z.1 M08
N810 G83 G99 Z–1.1 Q.134 R.1
  F7.8
N820 Y–1.0
N830 Y–1.5
N840 Y–2.0
N850 Y–2.5
N860 X4.0
N870 Y–2.0
N880 Y–1.5
N890 Y–1.0
N900 Y–.5
N910 G80 Z.1 M09
N920G91G28Z0
N930 M01
(Tool 6 = 3/8 inch diameter HSS
  Drill)
N940 T06 M6
N950 G90 G20 G80 G40 G49
N960 G00 G54 X3.25 Y–1.0
N970 S841 M3
N980 G43 Z1.0 H06
N990 Z.1 M08
N1000 G83 G98 Z–2.712 Q.25
  R–.4 F4.2
N1010 Y–2.0

N1020 X1.0
N1030 Y–1.0
N1040 G80 Z.1 M09
N1050 G91 G28 Z0.0
N1060 M01
(Tool 7 = 23/64 .3594 inch diam-
  eter HSS Drill)
N1070 T07 M6
N1080 G90 G20 G80 G40 G49
N1090 G00 G54 X2.125 Y–.75
N1100 S877 M3
N1110 G43 Z1.0 H07
N1120 Z.1 M08
N1130 G83 G99 Z–2.608 Q.24
  R–1.4 F4.4
N1140 Y–2.25
N1150 G80 Z.1 M09
N1160 G91 G28 Z0.0
N1170 M01
(Tool 8 = .3750 diameter HSS
  Reamer)
N1180 T08 M6
N1190 G90 G20 G80 G40 G49
N1200 G00 G54 X2.125 Y–.75
N1210 S387 M3
N1220 G43 Z1.0 H08
N1230 Z.1 M08
N1240 G85 G99 Z–2.75 R–1.4
  F5.8
N1250 Y–2.25
N1260 G80 Z.1 M09
N1270 G91 G28 Z0.0
N1280 M01
(Tool 9 = 5/8 inch diameter,
  HSS, 4-Flute End Mill)
N1290 T09 M6
N1300 G90 G20 G80 G40 G49
N1310 G00 G54 X3.25 Y–1.0
N1320 S505 M3
N1330 G43 Z1.0 H06
N1340 Z.1 M08

N1350 G82 G98 Z–.875 R–.4
  P200 F5.0
N1360 Y–2.0
N1370 X1.0
N1380 Y–1.0
N1390 G80 Z.1 M09
N1400 G91 G28 Z0.0
N1410 M01
(Tool 10 = ¼-20 Tap)
N1420 T010 M6
N1430 G90 G20 G80 G40 G49
N1440 G00 G54 X.25 Y–.5
N1450 S382 M3
N1460 G43 Z1.0 H10
N1470 Z.1 M08

N1480 G84 G99 Z–.75 R.1
  F19.0
N1490 Y–1.0
N1500 Y–1.5
N1510 Y–2.0
N1520 Y–2.5
N1530 X4.0
N1540 Y–2.0
N1550 Y–1.5
N1560 Y–1.0
N1570 Y–.5
N1580 G80 Z.1 M09
N1590 G91 G28 Z0.0
N1600 G28 X0.0 Y0.0
N1610 M30

CNC Machining Center Program Error Diagnosis Answers

Use the skills you have learned to identify the problems in the following program lines and program sections. You may refer to the text *Programming of CNC Machines*, Fourth Edition.

1. Use the CNC code to sketch a representation of the part being created by the following program. Refer to Figure 1-4.

O3001
N100 G90 G17 G20 G80 G49
(3/8 2FL ENDMILL)
N105 T1 M6
N110 G00 G54 X0.0 Y0.0
S1426 M3
N108 G43 Z.1 H01 M8
N110 G1 Z–.1 F6.33
N112 G41 D1 Y6.
N114 X1.0 Y7.0
N116 X1.5
N118 X2.5 Y6.0
N120 G3 X4.5 R1.0
N122 G1 X11.0 Y2.5
N124 Y1.

N126 X10.0 Y0.0
N128 X0.0
N129 G40 X–.2
N130 Z.1
N132 M5
N134 G91 G00 G28 Z0.0 M9
N136 G28 X0.0 Y0.0
N138 M30

2. Identify the missing information in the following program line.

>   N250 G02 X–.375 Y2.938

The arc center locations or Radius (R) designation are missing.

3. Identify the incorrect or missing information in the following program line.

>   N110 X–4.625 F12.

Missing G01 with Feedrate

4. Identify the incorrect or missing information in the following program line.

>   N25 S2500 M4

Incorrect Spindle direction code

5. Identify the missing information in the following program line.

>   N125 G83 G99 Z-1.13 R.1 F3.6

Missing pecking distance (Q value) in Canned Drilling Cycle

6. Identify the incorrect or missing information in the following program and/or subprogram.

```
O2010
N10 G90 G80 G20 G40 G49
M23
N15  G00  G54  X–1.25  Y.75
S1000 M03
N20 G43 Z1.0 H01 M08
N25 G81 G98 Z–.35 R.1 F6.0
N30 M98 P7
N35 G00 G80 X0.0
N40 M21
N45 G00 X–1.25 Y.75
N50 G81 G98 Z–.35 F6.0 R.1
N55 M98 P7
N60 G00 G80 X0.0 Y0.0
N65 M23
N70 M22
```

```
N75 G00 X–1.25 Y.75
N80 G81 G98 Z–.35 F6.0 R.1
N85 M98 P7
N90 G80 G00 X0.0 Y0.0
N95 M21
N100 G00 X–1.25 Y.75
N105 G81 G98 Z–.35 F6.0 R.1
N110 M98 P7
N115 G80 Z1.0 M09
N120 M23
N125  G91  G28  X0.0  Y0.0
Z0.0
N130 M30
```

Subprogram for Program 2010

        O2011
        N1 X–2.5
        N2 X–3.75
        N3 Y1.5
        N4 X–2.5
        N5 X–1.25
        N6 Y2.25
        N7 X–2.5
        N8 X–3.75
        N9 M30

Improper program # call, for subprogram in line 30, 55, 85, and 110. The correct information should be N30 M98 P2011 in each line.

There is an improper subprogram ending in subprogram #2011. Line N9 should read N9 M99.

7. Identify the incorrect or missing information in the following program line.

        N310 G01 G41 X–4.125 Y0.0

Missing cutter diameter compensation call (D#)

8. Identify the missing information in the following program line.

        N10 G90 G20 G80 G49

Missing cutter diameter compensation cancellation

9. Identify the missing information in the following program line.

        N35 G43 Z1.0

Missing tool height call (H number)

Index

Notes

Notes